Junk DNA

John Gaynor

ISBN: 979-8-90252-181-5 (Paperback)
ISBN: 979-8-90252-182-2 (Hardcover)
ISBN: 979-8-90252-183-9 (eBook)

Printed in the United States of America

Contents

Junk DNA

Book 1

Junk DNA is a DNA sequence that has no known biological function. Most organisms have some junk DNA in their genomes—mostly pseudogenes and fragments of transposons and viruses—but it is possible that some organisms have substantial amounts of junk DNA.

All protein-coding regions are generally considered to be functional elements in genomes. Additionally, non-protein coding regions such as genes for ribosomal RNA and transfer RNA, regulatory sequences, origins of replication, centromeres, telomeres, and scaffold attachment regions are considered functional elements. (Wikipedia)

Episode 1

Genodyne and the Neuro Helix

The Awakening Code: Genodyne and the Neuro Helix

In the quiet hills of eastern Tennessee, nestled between the Smoky Mountains and the winding Tennessee River, Genodyne—a small biotech lab—had been quietly working on something the world had long ignored: junk DNA.

Muncie Takes the Neuro Helix

I trusted Jack Dunkirk.

That was the first mistake—or the first requirement. I still wasn't sure which.

His lab sat half underground, concrete walls sweating Appalachian humidity, shelves cluttered with gene maps, folk charms, and outdated equipment he refused to throw away "on principle." Jack said the mountains had memory; he said DNA did too. I'd always chalked that up to his rebellious streak, the part of him that refused to believe anything as elegant as biology could contain waste.

"Junk DNA is a lie we told ourselves because we didn't know how to listen," he said, pacing like a preacher before a revival. "Ninety-eight percent of the genome just sitting there? No. That's not inefficiency. That's compression."

I crossed my arms. I am an engineer with military training. Every instinct I had was trained to distrust poetic explanations.

Still…I'd seen his data.

Jack stopped pacing and looked at me—*really* looked. "I wouldn't ask if I wasn't certain."

"Certain of what?" I asked.

"That it works."

There it was. The word that mattered.

He explained the Neuro Helix protocol with just enough rigor to satisfy me and just enough omission to unsettle me. Noninvasive. Temporary. Reversible. He'd already tested it on himself—*multiple times.*

"What's the catch?" I asked.

Jack smiled. "You don't get to decide what you notice afterward."

That should have been enough to stop me.

Instead, I sat.

The device—if that's what it was—rested lightly against my temples. Cool. Almost gentle. No needles. No restraints.

Jack's hands were steady as he made final adjustments, the hum of the equipment rising to a pitch vibrating behind my eyes.

"You can stop it anytime," he said.

I nodded. "I know."

I didn't mean the protocol. I meant life.

The moment it activated, nothing dramatic happened.

No pain. No light. No rush.

Then…

Perspective shifted.

Not outward. *Inward.*

It felt like standing inside my own thoughts and realizing the room was larger than I'd been using. Connections I'd always known were there—experience, intuition, memory—suddenly aligned, snapping into coherence like gears finding their teeth.

I wasn't hallucinating.

I was *integrating.*

Images surfaced unbidden: schematics I'd once glanced at and dismissed now unfolding into elegant solutions. Conversations from years ago replayed with new subtext, hidden variables emerging like annotations I'd somehow missed the first time.

Time didn't slow.

It layered.

"Jack," I said quietly because my voice felt important now, measured. "This isn't enhancement."

"No," he said, reverent. "It's access."

I understood systems, power grids, command structures. Failure cascades. What I was experiencing wasn't information overload; it was prioritization without effort. Noise fell away. Signal sharpened.

And beneath the exhilaration—because yes, there was exhilaration—I felt it.

Unease.

This wasn't something you could unknow.

My heartbeat was steady. My breathing calm. My mind…vast.

I could see implications unfolding years ahead, branching paths of consequence, ethical vectors bending under pressure. The engineer within me mapped outcomes automatically.

The former officer counted risks.

And the man…the man understood the danger immediately.

This changed who decided things.

"Jack," I said again, slower this time, "who else knows about this?"

He didn't answer right away.

That told me everything.

When the protocol ended, the room felt smaller. Dimmer. Like stepping back into a house after seeing the horizon from a ridge line. I was still myself—no confusion, no weakness—but something fundamental had shifted.

I was smarter.

Not faster. Not stronger.

Clearer.

Jack watched me closely. "You're quiet."

"I'm calculating," I said.

He laughed softly. "Good."

I stood, steady on my feet, and met his gaze. "You've changed the rules."

Jack's smile faded just a touch. "No. I've revealed them."

As I left the lab, one thought settled into place with unnerving certainty:

If this could be shared, it would reshape everything.

And if it couldn't…

Then the real danger wasn't what the Neuro Helix did to the mind.

It was who decided *who deserved it.*

Jack and Muncie met in undergrad school. Jack was starting a career in academia and Muncie, a career in the Air Force. Both pursued their careers separately but largely, and by chance, they reconnected, interested in renewing a friendship. Muncie retired from active duty as a colonel in 2018; after the untimely death of his wife, he became an officer at the Tennessee Valley Authority (TVA). During the 2020 layoff, Muncie laid himself off along with seventeen other engineers and five contracts with the TVA. With the untimely death of Sam Redkey, the number was reduced to sixteen. Muncie needed office space, and he knew where to find it. Quickly, he became heavily involved with the research at Genodyne.

Jack Dunkirk was a semiretired professor at the University of Tennessee that had formed Genodyne a separate company doing "special" research. As the five-thousand-square-foot building was largely vacant, Muncie made a deal with Jack and his then–newly formed company, Genodyne. Muncie would rent 2,500 square feet, convert it into office space, leaving Jack with 2,500 square feet for Genodyne. Once on board, Jack and Muncie would work collaboratively on current projects with junk DNA. The whole concept would need engineering at some point, and Muncie fell in love with the idea.

Together, they developed a **Quantum Resonance Scanner** and a **Quantum Resonance Frequency Generator** that could "ping" these dormant sequences.

One rainy evening in 2021, the scanner pulsed with a frequency never seen before. The junk DNA responded. It didn't just light up, it resonated, awakened.

Within days, Jack and Muncie began noticing changes in themselves: Their cognitive abilities expanded exponentially. They could visualize multidimensional equations, communicate telepathically, and even manipulate matter at the quantum level. But the most astonishing change was the emergence of a new organ—an ethereal structure in the brain they called the **Neuro Helix**, visible only through quantum imaging. This Neuro Helix allowed them to tap into a universal field of knowledge—what modern science will dub the **Consciousness Lattice**. This Lattice opens telepathic "channels," allowing communication links and knowledge signals at quantum frequencies and with other signals of the same resonance, allowing multispatial communication at faster-than-light (FTL) speeds. Both Jack and Muncie could telepathically receive these signals as well as the signals emitting from each other. In addition, both started having telepathic experiences. Both talked about the schematic diagrams that would come to them at odd hours. They started taking notes, and soon, a complete pattern of electronic products was assembled in the lab.

Genodyne's research revealed that the so-called junk DNA can be awakened at its unique resonant frequency, with each sequence possessing its own resonance. By harnessing quantum frequencies, the team demonstrated that DNA could be influenced across distances, enabling communication that travels faster than light. This awakening not only triggered cognitive leaps and new brainwave patterns but also led to the formation of the Neuro Helix, a novel organ in the brain. To facilitate these breakthroughs, Genodyne assembled the Quantum Scanner, Frequency Generator, and Resonance Generator. Their work culminated in the development of a prototype for FTL transport, though it remains unfinished and dangerous.

Kathy Gets an Offer

The morning spring mist rolled over the Tennessee hills like a living thing, soft and electric, curling around the trees and the low fences as if it didn't want to leave. I stood on the porch of the farmhouse, my hands wrapped around a warm mug of coffee, breathing in damp earth and lake air. Mornings were the hardest. Sam should have been here—grumbling about the chill, teasing me about my boots, already half-lost in whatever engineering problem lived rent-free in his head.

Instead, there was only memory.

Jefferson sat on the steps, thumbs flying over his iPhone, completely absorbed. He was thirteen years old, yet he was already moving faster than the world around him. I watched him and felt that familiar ache in my chest—the mix of pride and fear that came with raising him alone.

My phone buzzed in my hand: "Muncie Mathews."

I frowned slightly before answering. Muncie never called this early unless something mattered.

"Kathy," he said, skipping the small talk. "Any chance you could come by my office today?"

Something in his voice made me straighten. "Today?" I asked. "Is everything okay?"

"Yes…well, I think so. It's important. Could you be here around ten?"

"I can," I said, already running through logistics in my head. "I'll make it work."

"I appreciate it," he said. "And, Kathy…good to hear your voice."

When the call ended, I stared out toward the hills again. Muncie had been Sam's friend long before he'd been mine. TVA, late nights, impossible deadlines. After the accident, he'd quietly stepped in, helping with paperwork, making sure Jeff's school situation was solid, and checking in without hovering. I trusted him.

"C'mon, Jeff," I said, forcing brightness into my voice. "Time to go to school."

Jeff groaned but grabbed his backpack and followed me to the car.

By the time I pulled into the Genodyne parking lot, it was ten thirty. Late. My stomach tightened as I hurried inside, wondering if I'd already messed something up.

Jack and Muncie were waiting near the lobby.

"Kathy," Jack said with an easy smile. "Good to see you. How's Jeff?"

"He's doing well," I said. "One of only two kids in the district who plays both football and basketball." I smiled despite myself. "Grades are good. Could be better."

Jack laughed. "Sounds about right."

Before I could ask why I was really there, Muncie clapped his hands together. "All right, change of plans. We're taking you to lunch."

"Lunch?" I blinked. "Where?"

Muncie grinned. "Doesn't matter. What matters is the conversation."

That didn't help my nerves.

Texas Roadhouse was loud and familiar, the smell of grilled steak grounding me as we slid into a booth. Still, my pulse didn't slow.

Jack leaned forward, folding his hands. "Kathy, I'll get straight to it. Genodyne needs another scientist."

I raised an eyebrow. "I'm an organic farmer now, remember?"

"We remember everything," Jack said. "What we need is someone we trust. Someone smart. Someone with an open mind."

"And someone who can handle a challenge," Muncie added carefully.

Jack continued, his eyes steady on mine, "A new scientific challenge. One that'll make you smarter."

I shot a glance at Muncie, who frowned at Jack like he'd overstepped. But my heart had already started racing.

"You'd have to sign a nondisclosure agreement," Jack said. "We can't say much yet. But I've read your work—Purdue for graduate school, UT for undergrad. Physics major, genetics minor."

I nodded slowly. That combination always earned a reaction.

"Unusual," Jack said. "And exactly what we need. This project sits right at the intersection of both fields."

That was the key word: *Both.*

I felt something shift inside me—something I hadn't felt since Sam died. Curiosity. Purpose. The old thrill of standing on the edge of the unknown.

Research again. *Real* research.

"What about Jeff?" I asked, grounding myself. "I won't disappear on him."

"You won't," Muncie said immediately. "Part-time if you want it. Up to forty hours. You take him to school. You pick him up."

I studied their faces. These weren't men trying to trap me. These were men offering me a way back to science—*to myself*—without taking my son from me.

"I'm in," I said quietly, surprising even myself.

The papers came out. Pens were passed around. As I signed my name, I felt the weight of it—not fear, but momentum.

For the first time in a long while, the future didn't feel empty.

It felt alive.

First Day

Kathy enters the lab.

Muncie leads her to the bench. "We've experimented on ourselves—telepathy, cognition, schematics…all real. The Neuro Helix is the key. Some junk DNA responds, and some doesn't. Each has a unique resonance. We need you to find more frequencies. And help with FTL."

Kathy volunteers for the protocol, requesting an assistant.

Mira Levy joins six weeks later.

"So let's get busy" Jack said.

The Night Kathy Vanishes

I don't remember going to bed.

That fact comes to me later, like a splinter I can't quite dig out.

What I remember—what my mind insists on offering—is the sound of the refrigerator humming and the hard, cold bite of tile against my bare feet.

I come to standing in my kitchen at 2:07 a.m.

The digital clock on the microwave glows an unnatural blue—too bright for the dark house. For a moment, I just stand there, my hand resting on the counter, heart thudding as if I've run up a hill. My mouth tastes metallic. Dry.

"What the hell…?" I whisper.

The house is quiet—too quiet. No wind against the windows. No frogs from the lake. Just the low mechanical humming of appliances and the distant ticking of the old wall clock.

I look down.

I'm dressed. Not in pajamas—jeans, boots, my old TVA jacket folded neatly over the back of a chair like someone planned for me to wear it and changed their mind.

A ripple of unease crawls up my spine.

"Jeff?" I call softly, already moving toward the hallway.

His bedroom door is closed, just the way he likes it.

I open it slowly. Moonlight spills across the floor, catching the edge of his bed. Jeff is asleep, one arm flung over his pillow, mouth slightly open, breathing deep and steady.

Unharmed.

The knot in my chest loosens—but only a little.

I stand there longer than necessary, listening to him breathe, grounding myself with the proof that he's here, safe. Then I quietly close the door and lean against the hallway wall, feeling suddenly dizzy.

Something is wrong.

Back in the kitchen, I notice things I shouldn't.

My phone is on the table, face down. I always plug it in at night—*always*. The battery icon blinks red when I turn it over—6 percent. No missed calls. No notifications. No gaps in the timeline that make sense.

The back door is locked.

So is the front.

I check them twice.

There's no sign of forced entry. No broken glass. No mud on the floor. No smell of anything unfamiliar. Just my house. My kitchen. My life—slightly rearranged, like a room someone walked through without touching the furniture.

I press my fingers to my temple.

There's a headache there. Not sharp. Not painful. Just…pressure. As if someone pressed a thumb against the inside of my skull and forgot to lift it.

I try to remember the evening.

Dinner with Jeff. Homework at the table. A text from Muncie reminding me about paperwork tomorrow. I remember washing a plate. Turning off the light.

After that…

Nothing.

No fear. No dream. No interruption.

Just a blank stretch of time that my mind refuses to look at directly, like a word I know exists but can't utter.

I sink into a chair and rest my elbows on the table.

As a physicist, I know the difference between missing data and erased data.

This feels like erasure.

My skin prickles—not with panic, but with something colder. Clinical. Observational. The way I used to feel standing in front of a result that shouldn't exist.

Somewhere deep in my body, something knows I wasn't here.

I look at my hands and turn them over: no bruises, cuts, nor soreness. Nothing to explain the exhaustion that suddenly settles into my bones.

Whoever—or whatever—took me, if that's what happened, didn't want to leave a mark.

Didn't want me to remember.

I glance once more toward the hallway, toward Jeff's room, and a fierce protectiveness surges up, burning through the confusion.

Whatever this is, it didn't touch him.

That thought steadies me.

I clean the kitchen mechanically, wiping down a counter that's already spotless, moving objects back to where I believe they belong. By the time I crawl into bed, the sky outside the window is just beginning to pale.

As sleep finally drags me under, one thought repeats itself, slow and certain: Someone knows more about me than they should.

And they've already decided I'm worth taking.

Three Months Later

The lab is a hive of activity, with Mira Levy cataloguing DNA sequences, frequencies, and markers; so far, only four resonances and two markers have been identified. Remarkable changes are observed: aging reverses by 10 to 20 years, telomeres lengthen by 35%, and subjects experience surges in stamina, healing, and resistance to illness. IQ jumps by 40 to 60 points; and instant recall, multisensory perception, and telepathy are confirmed. Quantum manipulation becomes possible, as Muncie Mathews alters particle spin and Jack Dunkirk tunnels photons. The team launches a quantum AI project while progress on the FTL prototype continues, though failure remains a looming risk. Dormant DNA begins expressing new proteins, the Neuro Helix forms in the prefrontal cortex, and the QNFE device amplifies frequencies—still in the experimental stage.

Subjects report dreaming in geometry and unknown tongues, underscoring the profound effects of the protocol.

The Quantum Frequency Scanner pulses. Jack, Muncie, Kathy—Neuro Helix carriers—receive a transmission. Mira senses a warning.

Kathy screams. Signals flood the lab. Protocol 9 Alpha transmits this:

> **[BEGIN SECURE TRANSMISSION] – Ω-PRIME**
> **Origin:** P9-A | **Sector:** Interspace
> **Encryption Protocol:** Ω-Delta / Quantum Mesh | Layered Entropy Shield
>
> **>>> DATA STREAM INITIATED <<<**
> **VELOCITY:** 0.85 FTL | Quantum Drive: *Stable*
>
> Estimated **Nexus Gate** Entry in 6.4 FTL Months
> **Target:** Milky Way: SOL SYSTEM 7794 Alpha
> **Signal Integrity:** 91.2%
> **Timestamp:** [Pleiadian Galactic Standard]

4725.09.16 15:00
Mission Status:
Quantum Drive: Stable

ETA: Sol System 8.9 Sol Standard Months

Approach 4-degree declination

Shields: Not Active
Weapons: Not Active
Tactical Response: Phase-1 forward, sensing no threats
>>> END OF REPORT <<<

Mission Directive: Maintain current trajectory, ENTER Nexus Gate for Milky Way Sol System 7794 Alpha. On gate exit, proceed with a declination of 4 degrees, at .0002044 FTL mph; adjust accordingly. Prepare for census collection: Data initialized.
Response Notification: Identification of DNA Resonance SOL System 7794 Alpha.
The resonance is selective. Proceed with connection.
Acknowledge DNA Alpha patterns designate signal 31, 32, 33.
Detecting weak DNA Gamma pattern designate signal 41.
Message to designate signals; the Galaxy is dangerous.
This is Protocol 9 Alpha. Scouting mission, Pleiadian Naval Drone: Mapping first contact properties.

[END TRANSMISSION]

Protocol 9 Alpha divides the group: Dr. Jack Dunkirk sees a warning: Are they chosen or mere witnesses? Muncie Mathews doubts the signal's intent, urging caution and questioning who should cross this threshold. Dr. Kathy Redkey senses both invitation and danger, insisting on vigilance as answers remain hidden. In the midst of uncertainty, Mira breaks her silence: "I want to take the Neuro Helix protocol." But as ambitions rise and doubts deepen, the shadows of rivalry and threat begin to gather.

Quanta Dynamica is closing in.

Jack warns, "If they get our patents or Neuro Helix schematics, Genodyne is finished."

Kathy counters, "It's not just the tech. I vanished for one day, and someone accessed my research files. We need real protocols for personnel safety."

Muncie agrees, "I've contacted General Wallace for countersurveillance. We must encrypt blueprints, restrict lab access, and ensure that only verified staff touch the QNFE and QRG."

Jack pushes further, "Remote transmissions are vulnerable. If someone intercepts a Neuro Helix frequency, they could reverse-engineer everything. Quantum encryption and offline storage are essential."

Kathy adds, "Watch for insider threats as Quanta Dynamica may be recruiting from within. Background checks and security training are a must."

Jack concludes, "Let's draft a security protocol. Muncie, handle physical threats, Kathy, digital safeguards. I'll talk to the lawyers."

The threat continues to intensify.

Muncie updates, "General Wallace may secure military funding, but contracts bring complications."

Jack inflates the cautions: "Funding helps, but military involvement risks our intellectual property and research freedom. We're a research platform, not a manufacturer."

Kathy agrees, "We must protect our discoveries and ownership. I've noticed odd network activity. Could Quanta Dynamica be spying?"

Muncie confirms, "Surveillance cameras caught unfamiliar faces. Access controls are tighter and codes rotated, but we need a full audit."

Jack urges, "Check for digital breaches, Kathy. If Quanta Dynamica is probing, we must catch it early."

Kathy promises, "I'll alert everyone if I find anything. Until then, keep communications encrypted and avoid unsecured channels."

Muncie closes the conversation. "We'll keep General Wallace updated and ensure any funding agreement protects Genodyne."

General Douglass Wallace Visits Genodyne

Genodyne Lab, secure conference room

Muncie states, "General Wallace, Doug, good to see you and thanks for coming in. Our research and developed products are attracting unwanted attention. We're worried about both physical and digital threats, which Genodyne has been addressing. Let's start with the Quantum Scanner and work around to end with the Neuro Helix."

Muncie walks around the lab, explaining the quantum function of each piece of equipment and its relationship with the Neuro Helix. The general is astounded when he learns about the Neuro Helix and that Jack, Muncie, and Kathy have all been through the protocol, experiencing its benefits. He is somewhat skeptical yet amazed at the potential.

"General, now that you have seen everything, do you have any other questions?" Muncie inquired.

The general had some follow-up questions, and they were properly answered. "So what is your opinion on our research findings and if the military would be interested in funding some of it?"

General Douglass Wallace answers, "Funding? You're right to be concerned about security. Cutting-edge tech like yours is a prime target. What security measures do you currently have in place?"

Muncie explains, "We've installed biometric scanners and restricted access to sensitive areas. Surveillance is up. We're seeing signs of external monitoring—possibly from Quanta Dynamica."

"Muncie, I'm meeting with some colleagues very soon. I'm sure will have an interest and questions. Next, I will talk with them, and based on their interest, we will schedule another meeting. Are we good?"

All three of them nodded in agreement.

"Okay then, I will be in touch."

After the general left, Jack thinks he was an advocate on the idea of funding.

Muncie comments, "Look, he was put on the spot. He can't commit without the support of his peers and specific personnel. Give it some time. Let's see what interest he gets from his meeting."

Muncie and Sgt. William Roberts Security

Genodyne Lab Security Office (Muncie's Office)

"Hello, Bill, thanks for the callback. We've had a breach attempt on our offline storage. Our IP is possibly at risk, and we suspect someone may be trying to steal device blueprints and our IP," Muncie says.

Sgt. William Roberts responds, "Have you identified any suspects or patterns?"

Muncie answers, "Yes, possibly, a company called Quanta Dynamica. We're running audits and reviewing surveillance footage. We've also encrypted all research files and moved sensitive data to air-gapped systems."

"Good steps! Make sure to restrict lab access and monitor for unusual behavior among staff. Sometimes IP theft is an inside job. Do all your employees have badges?"

"No, they don't," Muncie admits. "I'm making notes right now. We're planning security training and updating NDAs. Any other recommendations?"

Sgt. William Roberts suggests, "Keep your legal team in the loop. If you spot anything suspicious, act fast. Disable access and preserve evidence. And don't hesitate to call outside for help if needed."

"Thanks, Bill. We'll stay vigilant," Muncie expresses his appreciation. "By the way, whatcha doing these days?"

Bill replies, "I do some consulting, running my security business—nothing dangerous.

Muncie inquires further, "So do you travel?"

"Yes."

“Okay, maybe a trip to Tennessee in the future?”

“I’d love that.”

Genodyne Lab, morning meeting

Tuesday

Two weeks later, Muncie’s cell phone rings.

“Hello, this is Muncie”

“Hi, Rangler, I’m going to be in town for some personal business this Saturday. Thought we could have breakfast or lunch. Are ya in?”

“Yeah, how about Ruby Tuesdays near the airport. Say, 10:00 a.m.”

“I’m in,” Muncie says.

Muncie calls a meeting.

“Mira, Kathy, in my office. Where’s Jack?”

Jack enters. “What’s wrong?”

Muncie starts talking. “Bill Roberts called. He used the code word *Rangler*—danger is coming.”

“Kathy, I’m worried about you and Jeff. Do you have somewhere safe to stay?”

Kathy admits, “I don’t know.”

Muncie instructs, “You and Jeff stay with me. Mira, what about you?”

Mira responds, “I have roommates.”

Muncie inquires, “How long has the newest one been there?”

Mira says, "Two months. His name is Howard Russel."

Muncie asks, "Can you stay with your parents?"

Mira answers, "Yes."

"Good. Kathy, does Jeff have a game Friday night?"

"Yes," Kathy affirms.

"Go to the game," Muncie instructs. "Jack, can you go with Kathy?"

Jack responds, "Sure."

Thursday 1:30 p.m., Genodyne HQ

Muncie's phone lit up with a text message: "FedEx special delivery for Rangler Mathews."

Muncie came to the front office and there was a FedEx driver with a letter that had to be signed. Muncie signed and took the letter to his office and opened it.

SATURDAY MORNING COMPROMISED

RESCHEDULE TO FRIDAY NIGHT 6:15 PM

At Landings Lounge dbl (don't be late)

Friday night, 6:15 p.m.

Muncie is a wreck all day long, thinking about the meeting—its cancellation. His stomach churns. He sits down and looks at the drink menu. As he looks up, he sees Bill and two other people.

"Okay, listen up," Bill says. "That attractive, intelligent-looking woman is Jesse. Don't get any funny ideas. Next to me is Steve. They are my agents. Quanta Dynamica is a shell company apparently working for a hostile Asian country. They have a nose for collecting intel any way they can, and sometimes people disappear. It is possible your whole office is compromised—your cell phone certainly is. You need a plan, and here it is. All your phones should be cleaned or replaced with new ones. That includes high-density security, which we can supply."

He continues, "Next, we sweep the office. You need some personal security for yourself, your officers, and key people in your organization."

Muncie interrupts. "A short time ago, Kathy Redkey is abducted for one day, and she remembers nothing."

Bill's eyes bug out of his head. "Muncie, what the hell are you into?"

"Well," Muncie says, "we have invented some equipment that sends communication at quantum FTL speeds. We have enabled individuals to communicate with enhanced telepathic methods. We are in communication with an unknown entity somewhere in deep space that states its mission is mapping first-contact properties. Hmm. I wonder if they have a real-estate license. We are developing an FTL device. Nothing too unusual—just your normal, everyday research."

Muncie looks at the three of them. They seem to be gasping for breath.

"You're serious, aren't you?" Bill asks.

"Yep."

"Holy shit. Your security needs just went stratospheric," Bill says. "Let me tell you about my immediate worry. It's Kathy Redkey. You said she was abducted—disappeared for two days. She comes back and can't remember anything. During that time, someone is accessing her records. I worry about her safety. She has a son named Jeff. He has a basketball game tonight. I'm afraid something will happen. Our owner, Jack Dunkirk, is with her, but I fear for him as well."

"What time's the game?" Bill asks.

"Seven thirty. Ends around nine."

"Muncie, we're going," Bill says. "Jesse, you're tailing us—watch for trouble. We'll pick up Jack and Kathy. Steve's alerted Rustler, code three. No one goes home tonight. Anyone else?"

"Kathy, Mira's on her own," Muncie replies.

"We don't have enough coverage for her. Let's move," Bill says.

Muncie pays the bill and steps out, scanning for a tail.

On the drive, Jesse checks in. "We're clean."

Inside, Bill lays out the rules. "Split up. Find Kathy and Jack. Stay hidden. Watch for anyone suspicious."

The game ends—Jeff's school wins. The crowd thins, tension rising. Jack and Kathy head toward the locker room. Kathy greets Larry St. John. Muncie and Bill follow, blending with the crowd.

Jesse whispers into her earpiece. "Bill, two zombies at the far exit. Eyes on Kathy and Jack."

"Stay in the shadows, Jesse," Bill says. "If they move toward Kathy and Jack, let me know immediately. Then follow them and close the distance as best you can without drawing attention."

"Copy that," Jesse says. "They're moving—heading your way, Bill."

"Code four, Jesse," Bill says, his voice tense. "Muncie, close in from the left. I'll intercept from the right."

Muncie moves quickly, catching up to Jack. "Kathy, Jack—stay close to me. Don't go anywhere alone."

Kathy scans the crowd, worried. "Where's Jeff? He should be out now."

"I see him coming out with the team," Jack says. "He's with a friend."

Suddenly, the two suspicious figures break from the crowd and rush toward Jeff and Kathy. Jeff's friend gets pushed out of the way by Muncie. Startled and defensive, he looks back at a scuffle.

"Jeff, Kathy—get behind me!" Bill shouts.

"Not today!" Muncie lunges forward, shoulder-tackling one of the attackers and pushing him into Jesse. She dispatches him with an elegant jujitsu move, slamming him to the ground. Stunned but aware, he gets up and runs away.

Jesse moves in as the second attacker ignores her and tries to grab Jeff. Bill intercepts, pulling Kathy and Jeff to safety. Jack shoves the attacker. Jesse body-slams him, striking his ribs and sending him to the floor, but he gets up and limps away. Jesse signals Bill with a universal "time to leave" motion.

Kathy hugs Jeff, shaken. "Thank you. Are you okay, Jeff?"

"I'm fine, Mom. What just happened?"

"You're all safe now," Jack says. "Bill, Muncie, Jesse—thank you."

Tom messages Jeff while Larry assesses the situation. "There is a woman and a man I've never seen protecting Kathy and Jeff," Larry says.

"Dad," Tom says, "the woman has a gun."

Tom, viewing the melee, prepares to go to Jeff's assistance. Larry sees Tom drop his gym bag and step toward the chaos. He intercepts him just as Tom breaks forward.

"Tom, not now," Larry says. "Watch and learn. There will be enough time if it goes south. We'll do it together. Let's see who wins. I think the woman and man calling the shots are the good guys. Did you get a message from Jeff?"

"Yes," Tom says.

"What does he say?"

"He says he's okay but doesn't know what's happening."

"Tom, look at Kathy. She seems relieved. I think the worst is over."

From a short distance away, Sidney Yorktown texts Jeff: *"Jeff, who are those people? What's going on? Are you okay?"*

Jeff replies, *I'm okay, but I have no idea. I'll get back to you.*

Sidney, Tom, and Larry watch as Kathy, Jeff, and the group are closely escorted out.

"Let's get everyone outside," Bill says. "We need to leave now."

"Threat neutralized," Jesse says.

"Yes," Kathy says. "I don't know what we'd do without y'all."

"Where's Steve?" Bill asks. He grabs the radio. "Whitehorse, this is BB. Come in!"

Nothing.

"Okay, no answer. I'm going to get him. Jesse, follow—not too close. If it's a shit show, code five and get them out of here fast."

Bill goes to the car and finds Steve on the ground, moaning. He has been knocked out while guarding the cars.

"Steve, are you okay?" Bill asks.

"Yeah. Listen—I don't feel good."

"Okay," Bill says, getting on the radio. "Iron Lady, Whitehorse is going to the barn. Code two actual. Start now."

"Roger that," Jesse replies. She turns to the group. "We are going to the cars now and meet Bill and Steve. Once there, I'm going to check the vehicles, and then we head to the farm. Bill and Steve are going to the hospital. Quietly now—I think the danger is over."

As Jesse and the group arrive at the vehicles, Bill says, "I have removed two trackers—one from each car. They must still be in the area. Dunkirk, where is your vehicle?"

"At the west end of the parking lot," Jack replies.

"Okay. Dunkirk, you're with me. Jesse, check the cars for explosives and one more time for a second tracker. Muncie, your truck should be clean. Jesse, give it a quick check. We're leaving your truck here. Muncie, get in with Jesse and proceed to the farm. Dunkirk, Steve, and I are going to the hospital. We'll see you later."

"Dunkirk," Jesse says, "give me your cell. Hey, kid," she calls to Jeff, "you have a cell—I saw it. Give it to me."

Before starting the trip, Jesse puts all the cell phones in a Faraday bag and zips it shut. The ride is quiet, with lots of turns. They reach the farm in seventy minutes. No problem—no tail.

Muncie gets a call from General Masters.

Saturday morning, 8:00 a.m.

Muncie's cell phone has just been cleaned. Jesse, with a mischievous grin, quips, "My, it was dirty! Where did you get all that porn?"

The room freezes.

Jesse's face turns beet red as she realizes Jeff is within earshot. Kathy Redkey glares at her with eyes that could melt steel.

Muncie frowns. "What the hell are you talking about?" Then, catching Jesse's sheepish look, he realizes it's a joke.

Jeff jumps in. "Oh, sorry, Muncie, I did that. Jesse, if you want the URLs, I have them alphabetized and color-coded!"

Kathy points at her son. "You! Up to your room! And don't come down until you've repented!"

Jeff laughs all the way up the stairs. He looks at Jesse and mouths, "Call me!" Kathy catches it.

"I saw that!" she yells.

Just then, Bill walks in, oblivious. "What's going on? Did someone win the lottery or just lose their dignity?"

"I'll tell you later," Jesse says. "Or maybe never."

Muncie finishes listening to a voice message. "That was General Masters. He says, 'Give him a call pronto.' You want to hear it?"

"Yes!" everyone shouts.

Muncie plays the message. "I was having lunch with some colleagues, and when I mentioned my Genodyne visit, the shit hit the fan. They were all over Genodyne and their cutting-edge research. Is it real? Did you meet

Dr. Redkey? What new projects are they working on? What kind of funding, and how much? Muncie, they want access now. Call me ASAP."

"So what do you say?" Muncie asks. "All in favor?"

"Aye!" the group responds.

"Muncie, with me. Outside," Jack says.

The room hushes.

"This whole project is getting out of hand," Jack says. "I feel like I'm losing control. You cannot make any deals with the military without my approval. I must be in every meeting."

"Okay, Jack. Not a problem," Muncie says. "How do you want to proceed?"

"First, we meet them here in Tennessee," Jack says. "I don't like the two of us being away from the office, knowing what just happened. And if anyone mentions URLs again, I'm bringing earplugs. One thing more—I'm going to make some organizational changes."

They walk back into the front room.

"Are you done?" Bill asks.

"Yes."

"Okay. Muncie, you and Steve come with me. We're going to pick up your truck. You drive it to the airport Ruby Tuesday's and leave it—text me when you're there. I'll pick you up."

He turns to Steve. "We are going to the airport, and you're going home to recuperate. I'm meeting two more agents at the airport today. You'll take our private jet home this afternoon. They are going to review and update your security. Once they're done, we will show you how to manage them."

He looks at Jesse. "You will hold down the fort. If you smell trouble, evacuate. In the meantime, there are steaks in the freezer—make a shopping list. Do we need charcoal? We'll leave here on Sunday, so get comfortable. Oh—and give your home addresses to Jesse. We will be conducting surveillance on your residences to make sure they are safe and secure."

"So, Bill, who's coming?" Jesse asks.

"Philly and Deuce," he says.

Jesse quietly smiles.

Episode 2

Philly and Deuce and the Keltthran Skirmish

Philly

Her real name is Patricia Holman, age twenty-nine. She is an ex-army buck sergeant and one of seven long-time employees of Roberts Security Corp. She was a truck driver in Afghanistan when her convoy was attacked. She grabbed her rifle and immediately returned fire. She did not show fear, as noted by her fellow soldiers. She commanded two privates to follow her. They flanked the enemy, killed three, and captured two. Sgt. Holman then called an air strike on a machine gun nest that had pinned down an accompanying squad of six.

She is only five feet, seven inches, 150 pounds of pure mischief, but she is very strong and agile. She is slightly heavy, attractive, and a college graduate with a major in history and minor in math. She is, by all outward indication, a geek, but is not one to keep to herself. She is sarcastic but not mean.

In the military, she gravitated toward communication and intelligence. After capturing the two enemy combatants, she ran fifty yards into a dwelling that she suspected of being a unit command center. She recovered a laptop, planning and personnel documents containing intel on an upcoming attack. That won her a bronze star for leadership and bravery. She mustered out after eight years due to boredom and the fact that the military would not move her to an intelligence position.

She met Bill in a bar shortly after. He was starting his own business, and she was an ideal candidate. The two established a friendship. She went on learning intelligence, communication protocols, equipment, security, surveillance, and improving marksmanship. Bill was providing personal security for several high-profile females, and she absolutely fit the position. About nine months after joining Roberts Security, her friend Jesse was hired.

Between the two of them, work was a lot of fun.

Deuce

He is thirty-three, an ex-army first sergeant, and divorced. He is also one of the very first employees at Roberts Security Corp. He was a combat veteran who was wounded in the Afghanistan War. He was then assigned to an intelligence company where, some say, Philly and Deuce met. Nobody but Bill knows what Deuce's real name is; he is just Deuce.

Some say he got his name because of a bar fight where he took down two combatants with one punch each—thus, the name Deuce. Rumor has it that Bill and Deuce were friends in the army and that Bill was in that fight. No information as to which side Bill was on. Deuce mustered out after ten years. He handles risky and other "special" assignments as he is knowledgeable with all kinds of weapons, communications, and security—both digital and personal.

He is six feet, four inches, and weighs 210 pounds. He is very imposing, intelligent, articulate, and attractive. He is not chatty but is amicable and very loyal to his friends. He was initially hired for an operation that remains classified and was rumored very dangerous. While he was hired for that engagement, his talents propelled him to many different assignments.

Roberts Security must be careful what assignments they use with Deuce as it was established early on that women frequently gravitate to him—which presented difficult and embarrassing situations for the company. Currently, he provides training for all new employees and handles any and all assignments.

When paired together, Philly and Deuce create a very deadly, intelligent, and adaptable team. They are best friends and frequently work together.

Sunday, 8:48 a.m.

As Jack is just waking up, Kathy and Jesse are already up preparing breakfast.

"Coffee?" Jesse asks.

"Yeah, smells good. What is it besides bacon?" Jack asks.

"Pancakes," Kathy answers.

The door opens, and two strangers enter—a male and a female.

"Philly!" Jesse yells. She puts the coffee pot down, runs over, and hugs her.

Kathy notices both are carrying sidearms. Then she notices Jesse is carrying as well.

"Let me introduce everyone," Jesse says.

Bill enters. "Is everyone introduced?" he asks.

"Just starting," Jesse remarks.

Muncie walks down the stairs. "What's all the commotion about?" he asks.

"Just introducing the new people," Bill says.

“Oh, hi, Deuce. Philly—we met yesterday,” Muncie says.

They both nod.

Jeff walks quietly down the stairs and nods to everyone.

“Good, we’re all here,” Jack says.

Bill turns to Deuce. “You requested a surveillance update,” Deuce reports. “Routine checks at assigned locations are made. There is minimal activity, except for two unidentified individuals at Muncie’s property. No other activity is reported. Muncie’s property has a gravel access road behind it where fresh tire tracks are noted.”

“Okay,” Bill says. “Here is what we’re going to do. Philly and Deuce, you’re going to plant cameras at the entrances to each house, except Mira’s. We will let her slide for now. At Muncie’s dwelling, we add extra cameras on the gravel road. Try to identify the access points at the intersection and place some there.”

He looks around at the group. “Jack, Muncie, Kathy, and Jeff—we will take you home later this afternoon. I would like one of my people to spend the night with each of you tonight. Is that okay?”

They nod.

“Okay,” Bill says. “Then it’s settled.”

“So here is what I want,” Bill continues. “Jesse, you are at Kathy’s.”

He looks at Kathy, who nods in agreement.

“Deuce, you are with Jack, and Philly, you are at Muncie’s.”

Jack stops and looks at Muncie and Kathy. “I’m copying Protocol Nine Alpha. Are you?”

"Yes," they answer.

Muncie and Kathy stop and sit down. No one moves.

Bill and the crew watch as Jack, Muncie, and Kathy close their eyes and begin what can best be described as meditation.

"Protocol Nine Alpha is entering Nexus Gate," Jack says.

Bill and the crew sit down, watching. They cannot believe what they are seeing or hearing.

There is a pause.

"I wonder if the Quantum Resonance Scanner copied that," Jack says. He grabs his cell phone.

"What are you doing, Jack?" Kathy asks.

"Just a minute, and I'll tell you," Jack replies.

He looks at the screen. "Okay—we got something."

"Jack!" Muncie and Kathy exclaim.

Jack beams at them. "I wrote an app for the equipment—all of it. This is the first live test."

He passes the phone to Muncie, while Kathy, Bill, Deuce, and the others look on.

[BEGIN SECURE TRANSMISSION] – Ω-PRIME
Origin: P9-A | **Sector:** Nexus Gate Entry
Encryption Protocol: Ω-Delta / Quantum Mesh | Layered Entropy Shield

>>> DATA STREAM INITIATED <<<

DISTANCE VECTOR: 2005 γ-units
VELOCITY: 0.20 FTL | Quantum Drive: *Stable*
Target: Milkey Way SOL SYSTEM 7794 Alpha
ETA: 1.5 Sol Standard Month | Approach 4-degree declination

MISSION STATUS

Activated **Nexus Gate** entry and exit protocols

Estimate **Nexus Gate** Entry in 25 FTL minutes

Activating **Quantum Override Channels** for Command Core reinforcement signal

Activating **Phase-3 Tactical Response**

>>> END OF REPORT <<<

Signal Integrity: 91.2%
Timestamp: [Pleiadian Galactic Standard] 4726.02.16
Hash Signature: SHA-512 →
A9F3C7E4D2B8F1A6C4E9D7B3A1F0C8E
6D4B2A9F3C7E4D2B8F1A6C4E9D7B3A1F0
C8E6D4B2A9F3C7E4D2B8F1A6C4E9D7B3

PRIORITY ALERT: OMEGA-LEVEL ESCALATION

Mission Directive: Maintain current trajectory. ENTER Nexus Gate for Milkey Way Sol System 7794 Alpha. On exit, proceed with 4-degree declination at .0002044 FTL 137,152 mph; adjust accordingly. Prepare contingency protocols for Gamma variance escalation, and population

categorization census PPC101.
BE ADVISED
Omega-Level Escalation: 2 Delta
Unknown vessel.
Moving toward Nexus Gate at high velocity.
Awaiting command sequence...
[END TRANSMISSION]

Bill takes one look at the communication and speaks. "Holy shit, Jack, this is some impressive work. Tell me—what did we just hear? What is this?"

Jack responds, "By the way, I have installed geosynchronous tracking in all equipment. If it's stolen, this app will find it. However, if it leaves Earth, it's a new game."

Later that day, as people are delivered home, Bill and Jack sit down to talk.

"Tell me about your equipment and what is going on," Bill says.

"There is a lot going on, Bill," Jack replies. "Let me talk about the big picture first—and briefly. Our equipment is tuned to quantum frequencies. Some of those frequencies correspond to the resonance of certain, very specific junk DNA. When the junk DNA experiences that signal, a whole bunch of things happen. One of the most important is the creation of the Neuro Helix in the brain. Once created, it facilitates telepathic quantum signals and other functions."

He continues, "Those signals can be broadcast to other telepaths with the same junk DNA signature—on quantum frequencies, not internet protocols. If those telepaths have a Neuro Helix, the signals are received and amplified. If not, some telepaths may or may not receive the signal."

Jack leans forward slightly. "Just imagine, Bill—if all your agents had a Neuro Helix, then it would be possible to message them very securely. But all telepaths on that frequency could receive it—not good for security. We are working on solutions."

He pauses. "However, we are finding that when junk DNA creates a Neuro Helix, other attributes are instantiated—things like ID, language, DNA history, heredity, and other traits. We are learning what we have to work with and developing software that would screen out only the resonant frequency you want. The software is coming along, but we have a lot more work to do—more knowledge to gain and more qualified people to bring in."

"What you are seeing here is a quantum communication with a Pleiadian drone conducting a scientific mission in our solar system," Jack says. "Now you know what we know. What's concerning is the escalation—we don't know much more. This is the second communication we've had."

"In the first communication, Protocol Nine Alpha recognizes our conscious levels as alpha and assigns us IDs of 31, 32, and 33. I believe it is the Neuro Helix communicating. We receive the transmission telepathically, and the quantum scanner records it digitally. It is now on file and retrievable."

Jack smiles slightly. "Let's see if my app works. I haven't had this much fun in a long time. This quantum AI we're developing is going to be out of this world, so to speak."

He gestures as he continues. "Think about this scenario: you are surrounded. One or more of your agents broadcasts a signal that puts all your combatants to sleep—you win. Our equipment can send, receive, and amplify quantum signals, including signals from the Neuro Helix."

Jack finishes his explanation, then points at Bill. "This is all raw data and must be kept top secret. You and your crew will have to sign NDAs with Genodyne."

Bill nods in agreement.

"Bill," Jack adds, "the question I keep asking myself is—are we weaponizing consciousness?"

It is then that Bill realizes what a deep—and possibly dangerous—thinker Jack Dunkirk is.

Monday

Muncie gets up at his usual time, turns on *Fox Morning News*, and hears them talking about a newly discovered comet called Three I Atlas. It is moving toward Earth at a high rate of speed—faster than normal for comets—and has some peculiar properties.

Muncie calls Jack immediately to wake him. "Jack, turn on Fox quickly. I think our spaceship has been discovered."

As it turns out, Jack is already up and watching the same thing.

Bill holds a conference call with his team. "We've got some important work to do," he says. "We need to evaluate and secure their data and intellectual property. We need to provide security for the people involved—especially Kathy and Jeff. The work and home premises must be secured, and there are young adolescents involved."

He continues, "I will be addressing this with Jack and Muncie today and tomorrow. We will have to sign NDAs before we leave. What we have here is a national security issue of a proportion I have never seen before, and we cannot ignore it. We need to get involved and help. Are there any questions?"

Everyone understands. They go to Genodyne and begin working on security.

Bill meets with Jack and Muncie and hammers out a security contract with Genodyne. Roberts Security is contracted to provide protection.

Tuesday, 2:00 p.m.

EST Meeting: Genodyne Management and Roberts Security

At the Tuesday meeting, Genodyne management and Roberts Security review security protocols. Roberts Security implements advanced military-grade communication security for all data files, phones, and research computers, replacing Starlink. Personal phones are upgraded to highly encrypted devices, and a reorganization separates engineering from research.

Only employees with six or more years of service qualify for the Neuro Helix protocol.

All eligible employees agree to undergo the Neuro Helix protocol, which is irreversible. Duty rotations are established, with Jesse providing system-administration security and prioritizing protection for Kathy and Jeff. The protocol process is experimental, and participants complete IQ tests as part of the evaluation.

About two weeks later

Friday, 11:30 a.m.

Jesse sits in the temporary security office. "Hey—anyone? I'm hearing something. Jack? Kathy?" she calls out.

Muncie is out to lunch.

Jack pokes his head out of the lab. "Jesse—get in here," he says.

A new transmission begins:

> **Speaking is Adm. Chryzen (Keltthran Empire).** "Unknown vessel, I am Admiral Chryzen of the Keltthran Empire. You have entered restricted space. Identify yourself, stop, and prepare to be boarded."

The Keltthran Cruisers are large, black, and covered with shiny shields and weapon turrets. Their primary weapons are large-pulse energy devices, augmented with pulse cannons but without kinetic weapons. The length typically spans **1,200–1,600 meters**. It appears to be a very formidable vessel.

Hull design: Streamlined yet heavily armored, optimized for deep-space travel and combat durability. Reinforced plating with energy-dispersing alloys to withstand plasma and ion cannon fire.

Protocol 9 Alpha (3I Atlas)

Timestamp: *[Pleiadian Galactic Standard] 4726.03.10*

> "Adm. Chryzen, we are Protocol 9 Alpha, a drone vessel of the Pleiadian Empire on a peaceful scientific mission. State your reason for boarding."
>
> Admiral Chryzen says, "Protocol 9 Alpha, you are a pirate vessel and have entered restricted space under the authority of the Keltthran Empire.

Prepare to be boarded for inspection. Failure to comply will be considered an act of hostility."

Protocol 9 Alpha replies, "Admiral Chryzen, we are not a pirate vessel. As we previously stated, we are on a peaceful scientific mission, and our mission parameters prohibit unauthorized boarding. Please clarify your intent and provide credentials. We are transmitting our peaceful mission status and request diplomatic protocol engagement."

Admiral Chryzen states, "Your refusal to comply with boarding procedures is noted. Prepare to be disabled if you do not surrender immediately."

Admiral Chryzen Raises shields and charges weapons.

Protocol 9 Alpha responds, "Hostile action detected. Defensive protocols engaged. We advise against escalation, Admiral. Any aggressive maneuver will be logged and reported to Interstellar Command Authorities."

Admiral Chryzen answers, "You have been warned. Energy weapon arrays are targeting your vessel. This is your final opportunity to comply."

Protocol 9 Alpha has this to say: "Defensive shields active, Weapons Arm, Target 442 Kilo, weapons systems. Command: 'Target acquired.' We will not initiate hostilities, Admiral, but will respond to any attack. Communication remains open for peaceful resolution."

Admiral Chryzen states, "Protocol 9 Alpha, you have refused lawful boarding. Prepare to be disabled."

The admiral's ship pivots, aligning its energy weapon arrays with Protocol 9 Alpha. Energy capacitors surge as the vessel fires a concentrated pulse phase disruptor energy, aiming to overload Protocol 9 Alpha's shields.

Protocol 9 Alpha expresses, "Hostile action detected. Defensive protocols engaged. Micro jump now!"

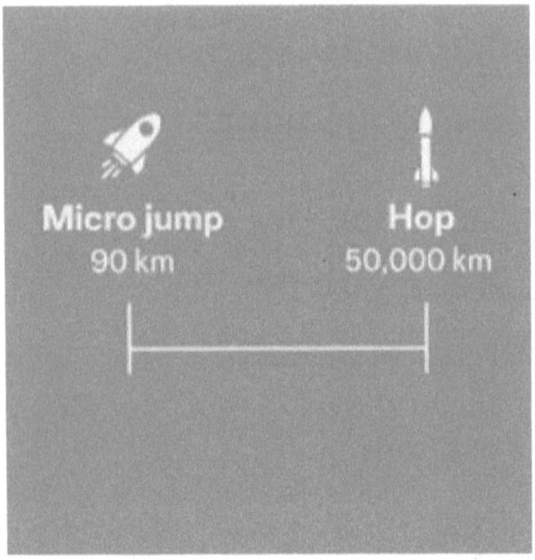

Protocol 9 Alpha's shields absorb a glancing blow due to the micro jump. Most of the energy blast missed, but signal integrity drops to 78 percent as the shields drop to 89 percent. The vessel executes a double-mike micro jump using quantum drive thrusters to evade a second volley. Population sensors continue to do their work.

After coming out of the double mike, Protocol 9 Alpha now designated 3I Atlas locks onto the admiral's vessel with its rail gun. Issuing a burst of kinetic energy, it fires a hypervelocity projectile, striking the admiral's ship along its starboard hull.

The kinetic impact partially disables several weapon arrays and disrupts the vessel's quantum mesh shielding. Admiral Chryzen's ship spins, partially losing attitude control and venting plasma from a damaged section.

"Admiral, this is Engineering, shields at 79% will be back online in two standard hours. Damage to communication array is fixable and is currently being repaired with an estimated completion of 2.5 standard hours. Lateral Control is returning and currently being repaired will be back online in 1.5 standard hours. The repairs for the plasma leak is just finishing. No casualties," reported Engineering from Admiral Chryzen's group.

This is from the Protocol 9 Alpha logs.

Timestamp: [Pleiadian Galactic Standard] 4726.03.10.14:22

"Computer log: Hostile Keltthran vessel is partially disabled. We suffered shield loss of 11% and communication loss of 22%. Regarding navigation, all other functions remain normal. Execute another defensive double-mike jump. Maintain trajectory, and continue mission. Increase velocity to .000234 FTL at 157,014 mph. When mission requirements are finished, initiate a hop to Vega Gate. Activate FTL Protocol Philo 2.0. Destination, Vega Prime."

Admiral Chryzen's voice is marked with suspicion. "Protocol 9 Alpha, what is a peaceful scientific vessel doing with armament?"

Protocol 9 Alpha answers, "Admiral, it is to protect our assets from bullies like the Keltthran. You should remember the history lessons from eight hundred years ago. Protocol 9, out."

The admiral's vessel, partially disabled, breaks off pursuit and initiates emergency repairs. As the admiral mulls over the event, he considers future conflicts and wonders what the Pleiadians are doing in the Milky Way's Sol System—scientific exploration?

Recognizing that the Keltthran vessel is partially disabled, Protocol Nine Alpha—now designated Three I Atlas—completes the mission and successfully initiates FTL Protocol Philo Two point zero, beginning with a "hop" to the Vega Gate. The drone remains on high alert, monitoring for pursuit or additional threats. Damage control teams assess minor shield degradation, but the vessel remains fully operational. The incident is logged as an Omega-level escalation involving an unprovoked attack. All communications are encrypted and transmitted to command for review.

> "Attention, Terran citizens of the third planet," the transmission states. "Be aware that your planet is rich with preformed four- and five-level consciousness. As such, it is at risk of attack and subjugation by the Keltthran Empire and other galactic bullies. We have also detected small populations of life-forms on your moon and fourth planet. You have allies: the Pleiadian Navy, the Andromeda Colonial Alliance, and the Vega Concordat. We will contact you on fourteen twenty megahertz and on secure quantum subchannel X-fifteen, which is already established with units Alpha thirty-one, thirty-two, thirty-three, and Gamma forty-one, forty-two, forty-three. Start planning your defense. The galaxy is dangerous."

Admiral Chryzen's vessel remains partially disabled but functional, venting plasma and operating with several systems offline. Emergency crews work to restore basic functions. The admiral's pride is wounded, but the crew's discipline holds. A report signal is sent, and the ship awaits completion of repairs while preparing for possible future encounters with Protocol Nine Alpha and, potentially, the Pleiadian Navy.

Sector Status

The skirmish is noted by nearby vessels and stations. Tensions in the sector rise, with both the Keltthran Empire and the Pleiadian Empire likely to escalate patrols and diplomatic communications. The encounter becomes a subject of analysis and speculation within both factions' command structures, as well as among Terran units.

Mission Completion and Crew Communication

This is from 3I Atlas: "Crew, prepare for departure," the command transmits. "Mission results: Earth is classified as a candidate for Keltthran Empire expansion and is marked as a Pleiadian first-contact candidate and a protectorate of the Pleiadian Empire. I recommend immediate escalation for Earth to Omega-level protocols. All data is secured and loaded into the Comm.Nav.D3 sensor. As we pass the large gas giant known as Jupiter, the sensor is released. Send the report to Pleiadian Navy headquarters at Vega Prime, status level: conflict."

Episode 3

31 Atlas Skirmish: The Empires React

Summary

The mission concludes with confirmation that Earth is rich in Alpha- and Gamma-level 4- and 5-preformed consciousness, marking it as a strategic target for the Keltthran Empire and other galactic bullies. The crew's findings are logged and transmitted to command, and preparations for potential invasion and defense protocols begin.

Significance

This outcome signals a major escalation in the sector, with Earth's unique consciousness profile drawing the attention of interstellar powers. The crew's professionalism and the vessel's advanced protocols ensure mission success and a safe return to command. Earth remains in a precarious position. With no defensive assets in the solar system, it is highly vulnerable to attack and subjugation. However, on the ground, Earth retains natural advantages, and harvesting level 4 and 5 consciousness may prove problematic for the Keltthran.

The Fallacy of Galactic Empires

Galactic empires boast millennia of stability, but beneath their rhetoric lies a history of exploiting weaker civilizations. These powers, complacent in their dominance, fail to adapt their tactics or technology, relying instead on past successes to maintain control. Their expertise lies not in innovation, but in identifying vulnerable civilizations ripe for conquest.

Jack's voice trembles as he articulates Earth's dire predicament. "We cannot afford hesitation," he says. "The Keltthran Empire will not wait for us to prepare. They will strike, exploiting our Gamma- and Alpha-level consciousness, unleashing psychic and technological devastation. Who speaks for Earth? Do we even have the authority to negotiate? Where are our defenses? And what about the Pleiadians? Have they honored their pledge of alliance, or are we alone?"

Muncie's internal storm mirrors the urgency of the moment. "We are exposed—like standing naked before a storm, waiting for the first strike," he says. "If we do not act, we will be swept away. We are utterly defenseless. The Pleiadians do exist—so why are we waiting? Let us reach out now. If we hesitate, we risk everything."

Kathy and Mira add further complexity, explaining that the Keltthran specialize in psychic suppression and that their technology surpasses Earth's by centuries. The stakes extend beyond Earth; the entire solar system could spiral into chaos if the planet falters. Urgent messages must be sent to the Pleiadians, the Andromeda Empire, and the Vega Concordat.

"If we stall, we are finished," Jack says. "The Keltthran—or anyone else—will see weakness and strike. We need to contact the Pleiadians, the Andromeda Empire, the Vega Concordat—everyone. Now."

"Wait too long, and we lose everything—reinforcements, allies, hope," Muncie adds.

"Then it is settled," Kathy says. "We act. Fast. For Earth. For everyone."

"Who tells our government?" Jack asks.

Muncie, feeling vulnerable and unsettled, steadies himself. "We need to send a message now—ask for help on the quantum subchannel," he says. "Bill, you are in."

Jesse hands the phone to Muncie as Bill, Philly, and Deuce join the call.

"Did you catch that?" Muncie asks.

"We did," Bill replies, his tone confident and positive. "We are ready for next steps."

"Yes," the others respond in unison.

"First, draft a memo for the generals," Bill says. "Show them what we are facing. Second, Genodyne upgrades. Third, contact the Pleiadians."

"Who else in the government?" Muncie asks. "When do we bring them in?"

"Meet the Pleiadians first," Bill says. "Ask them."

"Bill, let us list our military contacts," Muncie says.

"Muncie, Bill—invite the military," Jack says, with finality. "Start with General Wallace. Kathy and I will message the Pleiadians. Are we good?"

Three days later, the urgent message was sent to the Pleiadian Naval Command on Vega via the quantum subchannel X15.

Subject: Urgent Request for Assistance: Terran Defense Activation

To: Command of the Pleiadian Navy, Andromeda Colonial Alliance, Vega Concordat

Message body:

We are out of time.

Earth is exposed. Protocol 9 Alpha's report confirms it. The Keltthran have attacked your drone. We believe we are now a target.

We have nothing to stop them. No defenses. No hope.

Please—send your fleets.

Strengthen our psychic barriers.

Coordinate your rapid response.

Open secure channels: 1420 MHz, quantum subchannel X15.

We will fight. Train us!

If you hesitate, we fall.

Every second lost brings us closer to annihilation.

Please…mobilize now. Stand with us.

With gratitude,

Terran (Earth) Defense Command

Pleiadian Naval Station, Vega Prime

This is Pleiadian's Response to 3I Atlas:

"Captain Serenya, there's a conflict message from Protocol 9 Alpha exploration drone."

"Let me see it, Petty Officer."

Protocol 9 Alpha

[Pleiadian Galactic Standard] 4726.03.10 15:44; SPT

Status: Conflict Report

Location: Terran SOL System 7794 Alpha

Report:

Timestamp: [Pleiadian Galactic Standard] 4726.03.10.14:2

While on mission in the SOL System 7794 Alpha, we encountered a Keltthran Cruiser hailing us. We had become aware earlier that this vessel had been approaching from the Andromeda direction for quite some time.

Admiral Chryzen (from the Keltthran Empire) attempted to detain and board our vessel. He accused us of being pirates, demanded we stop and be searched, and then threatened to disable our drone before firing a pulse phase disruptor energy weapon.

Our course of action was in accordance with Pleiadian Galactic defensive response: 12,667.120.09.

Details:

I responded, denying that we are pirates and stating that we are a Pleiadian scientific drone on a peaceful mission, with mission parameters that do not permit boarding. I inform the admiral that we will not initiate hostilities but will return fire if fired upon.

The admiral responds with force. He fires a pulse-phase disruptor energy weapon that strikes our shields, causing an 11 percent drop in the main shield and damaging the communications array. The damage is not sufficient to force withdrawal.

We execute a double-mike jump and return fire with an L-fifty-three standard rail hypervelocity projectile. The round impacts the starboard side of the admiral's vessel, disrupting attitude control, degrading shields, and damaging several weapon arrays. The effect forces him to pause his pursuit and attack.

Recognizing his compromised position, I determine that we can complete the mission. We proceed and do so successfully.

Once the mission is complete, we update a Comm.Nav.D3 sensor and release it near the gas giant known as Jupiter. We then execute a hop to the Vega Gate and transition into hyperspace.

Estimated time of arrival at Vega Prime, Pleiadian Naval Station, is approximately thirteen FTL days.

See attachment for exact comms dialog.

Captain Serenya's voice sharpens with controlled fury. "They *fired* on our *drone*. Thank you, Petty Officer," she says. "This goes up the line." She is visibly angry.

She reads the Protocol 9 Alpha mission report. The third planet is overflowing with level 4 and 5 consciousness, along with a manageable number of Thetas. Seventy-two percent are progeny. The population is entirely unprotected, located in the Milky Way galaxy, and at risk of enslavement by the Keltthran Empire or another opportunistic power seeking an easy conquest. That detail alone will draw immediate attention. This requires action. She finalizes her recommendation, digitally bundles the report packages, and transmits them to Naval Command with an "urgent action needed" status.

Three days later, at 16:43 Vega Standard Time (VST), a signal comes through.

Captain Serenya, we have a quantum message from the Terran Sol System," the Petty Officer reports. "They are requesting our assistance."

"Give it to me, Sergeant," Serenya says.

As she reads the message, it becomes clear that immediate defensive support is required. She drafts her response, formally agreeing to the request for assistance. She proposes regular patrols in coordination with the Vega Concordat and sustained communication with the Andromeda Colonial Alliance. She also recommends filing a formal complaint with the Andromeda Galactic Legal Command and the Andromeda Colonial Alliance.

Pleiadian Response

Vega Prime, three days later

"Captain Serenya, urgent message from the Terran Sol System," the Petty Officer says.

Serenya snatches the transcript, her eyes blazing. There is no hesitation.

"We stand with Earth," she says. "Patrols begin immediately. File the complaints. The Keltthran will answer for this."

Six days later

Earth, East Tennessee, Genodyne Offices

Jesse and Kathy sit with cups of coffee.

"Feel it?" Kathy asks.

"Yes," Jesse says. "I think it is a message."

"Let's get to the lab—see who is there," Kathy says.

They enter the lab to find everyone already assembled. Jesse's phone buzzes.

"It is Bill and Philly," Jesse says. "Where is Deuce?"

"On assignment," comes the reply.

Jack looks up from the console. "We are receiving a message from the Pleiadians," he says.

To: Terran (Earth) Defense Command
From: Pleiadian Naval Command, Vega Prime

Subject: Re: Urgent Request for Assistance—Earth Defense Activation

Message Body:

Terrán Defense Command,

We acknowledge receipt of your urgent transmission regarding the imminent threat posed by the Keltthran Empire. The Pleiadian Navy recognizes the strategic significance of your planet and the vulnerability of its population to psychic and technological assault. Your request for assistance has been reviewed at the highest levels of our command.

In response, the Pleiadian Navy, in coordination with the Vega Concordat and the Andromeda Colonial Alliance, is prepared to mobilize immediate support for Earth's defense. Our actions will include:

- Deployment of advanced stealth quantum sensors, including Comm.Nav.D4, and other defensive countermeasures to reinforce planetary defenses.
- Strengthening of psychic barriers to protect Earth's population from suppression and manipulation.
- Initiation of joint tactical maneuvers and rapid-response patrols within the Sol System, focusing on key jump gates and vulnerable sectors.

- Establishment of secure communication channels via 1420 MHz and Quantum Subchannel X15, X27 protocols for real-time coordination.
- Diplomatic engagement with allied fleets and legal authorities to file formal complaints against the Keltthran Empire and seek broader interstellar support.

We urge Terran Command to facilitate direct liaison with our diplomatic and tactical teams to ensure seamless integration of defense protocols. The Pleiadian Navy stands ready to assist, regardless of the outcome of preliminary talks, and will provide protection as necessary.

Time is critical. We commend your swift action and professionalism and reaffirm our commitment to safeguarding Earth and maintaining stability in the sector.

A formal complaint has been filed with the Andromeda High Council; a copy is attached.

As I understand it, the Keltthran have replied.

With respect and urgency,

Captain Serenya

Pleiadian Naval Command, Vega Prime

This is the Keltthran's response to the Pleiadian complaint.

To: Pleiadian Naval Command, Vega Prime

From: Keltthran Empire High Command

Subject: Re: Formal Complaint—Alleged Attack on Pleiadian Science Drone

Message Body:

Pleiadian Command,

Your message has been received and noted. The Keltthran Empire finds your complaint both amusing and misguided. The Sol System, while not officially under our jurisdiction, lies well within our sphere of influence, and our patrols will continue as we see fit. Your so-called "peaceful scientific drone" was observed conducting activities that raised legitimate suspicions. Admiral Chryzen acted with full authority and in accordance with Keltthran protocols.

Accusations of piracy are not made lightly. If your vessel had nothing to hide, it would have submitted to inspection. Instead, it chose to resist, escalate, and fire upon a Keltthran cruiser. Your invocation of "Intergalactic Law" is irrelevant; the Keltthran Empire enforces its own standards, and our response was entirely justified.

If the Pleiadian Navy wishes to avoid further incidents, it would be wise to respect the realities of power in this sector. We suggest you reconsider your presence in contested regions and refrain from provocative actions that invite consequences.

The Keltthran Empire will not be lectured by lesser powers. Your complaint is dismissed.

With due superiority,

High Command

Keltthran Empire

Subject: Re: Invitation to Earth—Preliminary Meeting on Defensive Actions, January 10 at 10:00 a.m., Earth time

To: Command of the Pleiadian Navy, Vega Concordat, Andromeda Colonial Alliance

Message Body:

Esteemed Allies,

Taking into consideration the recent developments and the imminent threat posed by the Keltthran Empire, the Terran Defense Command respectfully invites the representatives of the Pleiadian Navy to attend a preliminary meeting on Earth. The purpose of this meeting is to discuss and coordinate immediate defensive actions, including reinforcement of psychic barriers, deployment of quantum countermeasures, and tactical maneuvers within the Sol System.

Your advanced expertise in quantum technology and empathic defense systems is critical in safeguarding Earth's population and maintaining interstellar stability. We propose to establish secure communication channels and direct liaison with your diplomatic and tactical teams to ensure the seamless integration of defense protocols.

Time is of the essence. Please let us know when the earliest earth date you will be available. We look forward to your prompt response and to working together in defense of our shared interests. We will be inviting a small (three-person) military contingent. At present, we don't know if they will be in attendance.

The meeting place is a working office space for Genodyne, and space is limited. We have a small area available for nineteen people. We will have seven.

With gratitude and urgency,

Terran (Earth) Defense Command

Subject: Re: Invitation to Earth—Preliminary Meeting on Defensive Actions

To: Terran (Earth) Defense Command

Message Body:

Terran Defense Command,

We acknowledge receipt of your invitation regarding the imminent threat posed by the Keltthran Empire to Earth. The Pleiadian Navy, in coordination with the Vega Concordat and the Andromeda Colonial Alliance, is prepared to attend a preliminary meeting on Earth to discuss and coordinate immediate defensive actions.

Our teams are ready to collaborate on the reinforcement of psychic barriers, deployment of quantum countermeasures, and tactical maneuvers within the Sol System. We will also establish secure communication channels and direct liaison with your diplomatic and tactical representatives to ensure seamless integration of defense protocols.

We confirm our attendance at the meeting scheduled for on Earth.

Muncie says, "Oh great! We've got buy-in from the Pleiadians, and the American military are not in the loop. So, Jack, here is what I'm doing. I'm trying to contact General Wallace, and I'm on permanent hold. Bill is trying also with some of his contacts and with the same results."

Jack responds, "The meeting on the eleventh is a go with or without the military. Why don't we meet with them earlier—say, on the fourth or fifth and clue them in? We could do this on a Zoom call, or we could meet somewhere else."

An hour later, Muncie invades Jack's office.

"Jack," he begins, "we're on the calendar for January 5 in the Pentagon at 10:00 a.m."

The Pentagon

January 5, 10:00 a.m.

Jack and Muncie enter the main area, passing through the security inspection and into the waiting room. The room is light but drab; the muted paint inspires boredom rather than patriotism. After a fifteen-minute wait, a captain arrives to verify their credentials and escort them to General Wallace's spacious office. Inside are Admiral Sara Grant and General David McDougal.

"Greetings, Jack and Muncie. Please have a seat," General Wallace says. Jack feels nervous after the introductions.

Jack begins, "We have brought transcripts and recordings on a thumb drive for a presentation—and an invitation. Genodyne has developed quantum technology that allows amplification of telepathic communication and reception of quantum signals on a frequency labeled X15. These are recordings of what was called Protocol Nine Alpha, now identified as 3I Atlas. The audio was received and recorded about three weeks ago. The time is galactic standard, and I do not understand the conversion to Earth time. Here it is." He hands over the transcripts and the thumb drive.

"I know this is blunt, and I am not apologizing. I do not expect you to believe any of this, but we have an appointment with the Pleiadians on Saturday, January 10, at 10:00 a.m., in the Genodyne Offices to discuss the defense of planet Earth. I do not know if you are aware, but 3I Atlas was attacked near our solar system. My quantum technology recorded the event. We have the transcript of the battle. The transcripts are on the thumb drive—both audio and text. We have limited space."

Admiral Grant speaks up, "Well, I, for one, would like to meet them. I will clear my calendar." The other two nod in agreement.

"Ya know, Muncie, this is hard to believe," General Wallace says. "If I hadn't known you this long, I would laugh you out of town."

Muncie thinks she looks familiar—something about Admiral Grant stirs a memory.

As they leave the general's office, Admiral Grant is holding an impromptu meeting.

"Generals," she says, "I want to see this quantum equipment. Let us open our minds to possibilities. If their equipment is any good—or has promise—the USA needs to benefit."

Episode 4

Pleiadian Saga: First Contact with Eos

Genodyne Office near Knoxville, TN

January 10, 2026, 10:00 a.m.

Jack Dunkirk opens the meeting with introductions and a statement of purpose. He invites all to view the equipment and informs them that an explanation of the Neuro Helix will follow at the end.

Admiral Lumeris begins, "I am Admiral Lumeris of the Royal Pleiadian Fleet. I have served in the Navy for 1,050 of your years. My age is around 1,525 Earth years. The Pleiades, our home world, is a star cluster located in the constellation Taurus. This group of stars, also known as the 'Seven Sisters,' can be seen from Earth and has been studied since ancient times.[1] The Pleiadians are your closest DNA kin in the known universe.[2] You are our ancient progeny. We are human, not humanoid.[3] There are many myths from different cultures about Pleiades, but we will not go into them today.[4]

"Our recent mission brings us here to Earth, purposely, to address an old problem in the galaxies: the defense of the Pleiadian species and others in need of protection. My second battle, some eight hundred years ago, involved the Keltthran Empire in a decisive two-day fight. One of our patrols was ambushed near the Vega Alliance. We were overwhelmed by two cruisers and four destroyers. Our psychic battle abilities had just been updated but never tested; this was the test. We wrestled with and telepathically neutralized three destroyers in twelve hours of battle, leaving our cruiser damaged. From the beginning, we hit their cruisers and destroyers with rail guns, surprising them. They kept fighting. We fired another barrage of rail projectiles at the cruisers—most missed—and our cruiser suffered further damage. We finished them off with a deadly psyop attack as a last resort, the first of its kind ever recorded. Their cruisers became devoid of life. The destroyers registered only 40 percent life signs, though one remained largely functional except for its weapons. They turned and limped back toward Andromeda. The cruiser was left as a warning. It was the psyop attack they feared the most. On that cruiser were four admirals, six politicians, and twenty high-level bureaucrats and political fundraisers—all dead. That got reported, and the galaxy took note. Pleiades was now an apex power.

"Throughout time, peaceful Pleiadian progeny have been enslaved, murdered, and assaulted by other arrogant species. As a result, we ran, hid, and spread through three galaxies, finally settling where we felt safe. For years, we ignored the awful events occurring elsewhere. Our high attainment of consciousness makes us a target, yet we remain peaceful. As I've stated, Pleiadian consciousness has long attracted the attention of galactic bullies and powerful empires. Why, we asked? We were then—and are now—largely pacifists.

"So we launched a stealthy, systematic investigation over time, in three galaxies—Andromeda, Vega, and the Milky Way—and in three of our own solar systems. Our research revealed that these galactic powers were capturing, enslaving, and reconditioning the consciousness of our progeny into extremely powerful warriors. These warriors excelled at the task, which attracted even more attention. This enslaving is akin to rape in your society. Over time, we became more militant.

"In our own solar system, no discord or signs of these problems occurred, though we noticed small populations of Pleiadians mysteriously disappearing. In Andromeda, our progeny was diminished by 78%. In the Milky Way, about 30%, and in Vega, 12%. This was horrifying. We launched operations to identify and rescue our progeny from Andromeda, relocating the survivors to two planets of the Seven Sisters."

Admiral Lumeris sits, and Ambassador Lyara stands.

"I am Ambassador Lyara of the Pleiadian High Council, and I am the oldest member of this team," she says. "I will explain the next phase. This effort took a long time, and we thought we had recovered all our progeny. On our last trip home with the final batch of refugees, our ship was attacked by a Keltthran cruiser from the Andromeda Galaxy. We lost that battle and all our progeny, but we fought fiercely. It was then we discovered how viciously smart and brave Pleiadian warriors could fight. We also noted our ability to defeat an enemy with a telepathic attack. This technique defended our ship, though it was not part of our armament at the time. Today, it is part of our defensive and offensive weapons. It was frightening and enlightening. It was our telepathic attack that drew the attention of all the bullies in the galaxy, even though we lost that battle.

"There was great debate over whether we should form a defense armada or rely on diplomacy. Recognizing the Keltthran Empire's arrogance, we decided to do both. The diplomatic war was a stalling tactic to give us time—time to create a navy capable of defending our home world. In Vega, the Concordat realized our vulnerability and assisted us. They have been excellent allies over time. Meanwhile, our scientists created advanced technology, which we shared freely with the Vega Concordat Navy. We established a naval base and scientific research center on Vega Prime over

2,400 years ago. Today, we have three naval bases, along with cultural, diplomatic, and trade centers in Vega and Pleiades. We share defensive responsibilities with the Concordat. Our diplomats operate in all three galaxies, and we are exploring new missions in twenty new solar systems. That brings us to Earth."

Ambassador Lyara sits, and Ambassador Zephar rises.

"I am Ambassador Zephar," he begins. "I am 960 years old, a reserve captain in the Pleiadian Navy, and the youngest member of the team. Earth, in Sol System 7794 Alpha, is the planet with the richest consciousness of the three galaxies. It has a population of 8.27 billion,[5] of which approximately 4.87 billion—or 58%—exhibit level 4 and level 5 consciousness. That makes your planet a prime apex candidate for conquering. Pleiades will protect the Sol System, specifically the third planet, from any hostile activity. Our combined fleet, diplomatic, and military assets are considerable and well known for their competence, fairness, and compatibility with new civilizations.

"Your civilization is like our own, and here we will be working with our progeny. Of the 8.27 billion population, about 72%—or 5.95 billion—share similar or identical DNA with the current Pleiades population. Not all are level 4 or 5 consciousness. We Pleiadians would like to discuss how you accomplished this advancement. Our recent census, conducted by 3I Atlas—a Pleiadian drone operating across two galaxies—was intercepted and attacked by a Keltthran cruiser in interstellar space near the outer reaches of Sol. This violated interstellar law.

"Our proposal is as follows: Pleiades will provide Earth with protection. We currently provide naval assets for your defense. We are building communications, supply, military assets, and defense substations in your solar system. We will share technology and training to help Earth defend itself and become self-sufficient. We would also like to open cultural, diplomatic, military, trade, and civic centers in Sol. We ask for your cooperation. The first step is the defense of Sol while simultaneously opening diplomatic relations."

General Wallace asks, "What if we object to your protection and the construction of assets on moons in our solar system?"

Ambassador Lyara responds, "Good question, General. There are things you should know about Galactic Law. Pleiades has filed a notification writ with the Interstellar Legal Counsel informing them that Sol System 7794 Alpha is under our protection. This is a legal and defensive move for your benefit. It prevents other entities, such as the Keltthran, from doing the same. This does not imply ownership or that you grant us any rights to your minerals, trade, or other privileges. Those belong to the planets in your solar system unless conquered. Earth may claim those rights, but it is not required. By location and law—what we call the 'Under the Sun Law'—they belong to the existing solar system, Sol.

"The notification informs all galactic organizations that you now have protection and are not defenseless. It lasts for one standard century. At any time, Earth may remove this protection; we will assist with the necessary filings. Earth is politically and culturally fragmented. You need to speak as a single entity to the Galactic and Interstellar Council. That means you are a class 1 or 2 species. As such, you are recognized by galactic and interstellar law as too primitive to defend your territory. Again, we are not here to conquer but to help. By galactic and interstellar law, Earth sharing DNA with Pleiades gives us an unassailable right to protect this solar system. This is called the By All Rights law. Earth may protest; if you do, Pleiades will claim 'By All Rights' and countersue. By the way, Pleiades provides classes and certification in galactic and interstellar law."

Jack addresses the Pleiadian delegation: Jack rises hesitantly, his voice resonant. "So," he says, slightly irritated, "because Earth is comprised of multiple cultures with no single voice, you claim Earth is unworthy of recognition?" This draws immediate attention. "Well, I say there is a new paradigm: the Neuro Helix. Your ears are deaf and your eyes blinded by old paradigms. Genodyne holds the keys to this paradigm—a new consciousness, and by implication, dominance, good or bad. Knowledge is structured in consciousness and moves at faster-than-light speeds. Consciousness carries insight, awareness, energy, intelligence, and the keys to unlock knowledge and dominate (good or bad). Ignoring this new paradigm—and the Neuro Helix—treats Earth as a subject, not an equal, at great risk."

The Pleiades respond: Ambassador Lyara rises slowly, a look of recognition crossing her features. She speaks softly. "Dr. Dunkirk, there are, right now, more ears and eyes sensing this dialogue and the truths you have spoken. Telepathically, great thinkers on Pleiades are observing the wisdom in your words." She moves closer, looking Jack in the eyes, then pauses. "We agree with your analysis. Be aware: Pleiades is not a dominion of aggression or subjugation. When we commit to the fight for good, we are intrepid, aggressive, and dominant. Earth—a planet with multiple cultures, politics, and no effective single voice—presents an opportunity for more powerful, aggressive cultures to conquer it. The galaxy is dangerous. In our collective consciousness, Genodyne holds keys, but not the defenses to secure them safely. Pleiades does. What we offer is a partnership, not subjugation."

A great silence falls as everyone contemplates the truths and the interplay of good and evil. Hesitantly, Kathy Redkey raises her voice. "Another question, please. Ambassador, you have mentioned consciousness. What is it? How do you define it?"

Ambassador Lyara, relieved, replies, "Good question, Dr. Redkey. It is simple to ask but complex to answer. Consciousness is a signal, a resonant frequency. When we raise consciousness, that resonant frequency increases. The brain allows consciousness to interact with reality.[6] It is fundamental to our reality on a quantum level. Consciousness can create matter and meaning. Meaning creates reality based on resonant frequencies. Words are symbols that create meaning. Meaning is knowledge or information. Information has structure, and structure affects reality. Meaning also creates patterns, which affect probability. These resonant frequencies can amplify to destruction or nullify to disappearance. Consciousness can be used as a weapon to attack or defend, but it requires specific training. The existence of the Neuro Helix, as explained by Dr. Dunkirk, is of great interest to Pleiades and other species we know."

Genodyne

Saturday, January 10, 1:46 p.m.

Admiral Lumeris looked at his watch and spoke something into it. Looking around, it was all quiet as the four Genodyne personnel stared at each other. Jack bellowed out, "Everyone upstairs to Lab 3! Follow me!" He crashed into the door and ran down the hall to the stairs; everyone followed. Muncie looked at Grant—they were having a quick tête-à-tête in the corner.

Grant asked, "Admiral, I saw you look at your communicator. What is happening?"

"Sara, there is a conflict at the EOS Gate. Keltthran ships and a Terran Destroyer. I sent our Destroyer D229 to investigate. The Destroyer should be there about now."

As they crowded into the lab, the quantum scanner picked up the EOS Gate confrontation.

First Contact: The Battle Over EOS Gate

The Rift Opening

Saturday, January 10, 1:46 p.m.

The hum of the Phase Rift sensors was the only sound on the command deck of the Terran cruiser *Atlas*. Admiral O'Brian stood like a statue, his silhouette framed by the cold glow of tactical holos. Outside, the EOS Gate shimmered.

"Unidentified Rift signature," the tactical officer whispered, voice cracking. "It is...large."

O'Brian's jaw tightened. "Bring up visual. Full spectrum."

The viewport flickered, and space tore open like a wound. A violet rift pulsed, its edges writhing with energy. From its depths emerged predators—sleek hulls bristling with plasma turrets, their green engine flares casting eerie shadows. The lead ship dwarfed the *Atlas*, a Keltthran Cruiser, flanked by two Destroyers like pack-hunting wolves.

"Sir..." The officer swallowed hard. "Hull composition matches Keltthran Cruiser."

O'Brian allowed himself a grim smile. "So it is." He sighed.

The Voice of Emoire

A guttural voice crackled through the comms, translated by the *Atlas*'s linguistic AI:

"By decree of the High Imperator, this world is claimed by the Keltthran Empire. Submit or be erased."

The words hung in the air like a death sentence. O'Brian felt the weight of history pressing down—Terrans' first true contact, and it was hostile.

"Raise shields, arm quantum torpedoes, and rail guns," he ordered. "Target their shields and weapons."

"Admiral, I'm getting another signature." As the vessel decloaked, Lt. Aer-ti exclaimed, "It's Pleiadian! It looks like a Vanguard-class Destroyer."

In an instant, the Pleiadian Destroyer decloaked and appeared with its shields activated. Its Tachyon Torpedo Launchers and Gravitic Railguns were armed and targeted at the three Keltthran fleet vessels. The *Pleiadean Destroyer D229*, its shields shimmering with a blue-white radiance, instantly signaled its readiness for battle. The hull was sleek and angular, optimized for rapid maneuvers and micro-jumps that allowed it to evade even the most precise enemy fire. Along its flanks, advanced Tachyon Torpedo Launchers and Gravitic Railguns were mounted, capable of disabling multiple targets in quick succession. Its tactical systems were engineered for rapid and precise strikes—to disable and destroy enemy ships.

"This is Captain Theissan of the Pleiadian Royal Navy. This territory is a Pleiadian Protectorate. Stand down and leave or be terminated."

The Keltthran Destroyers re-targeted their weapons on the lone Pleiadian Destroyer.

A guttural voice crackled through the comms. "This is Admiral Zexztar of the Keltthran Empire. Pleiades has no claim on this territory. Be prepared to die. FIRE!"

The Pleiadian Destroyer instantly executed a micro-jump to avoid a direct hit, then simultaneously fired one Rail Gun projectile at the large target and one at each of the two destroyers. The projectiles hit their targets, partially disabling shields, weapons, and venting plasma.

"Three partially disabled enemy vessels, all venting plasma. It's a good day in the neighborhood," said Captain Theissan.

This activity initiated the Telepathic Battle Group, attacking the minds of the Keltthran Admiral and his crew. Secondary attacks disrupted the control of the Cruisers' weapons and navigation, turning the ships away from the scene and toward the jump gate.

"Keltthran Destroyers and Admiral Zexztar, I have left your engines untouched so you can remove yourselves from this solar system. Otherwise, I put a Tachyon Torpedo up your collective asses, and you and your destroyers can go to hell. Thirty minutes to move or die. Time is ticking. Give my regards to the High Imperator," he spoke mockingly.

The Destroyers and Cruiser turned slowly and moved away. As the fleet moved out, Captain Theissan heard something the Keltthran Admiral was saying—it sounded like he was barfing.

"Which one of you," speaking to the Telepathic Battle Group, "caused the admiral to barf?" Two hands went up—one Pleiadian, one Terran.

"Pleiadian Destroyer D229, this is Admiral O'Brian of the *Atlas*, Terran Space Fleet. Where were you and how did you know?"

"We are in your system on a diplomatic mission. Keep all this quiet, Admiral, as your planet doesn't know yet. This solar system is now part of the Pleiadian Protectorate. This is not a hostile takeover, Admiral, but a cooperative defensive initiative to protect your solar system from activities like this. We will see a lot of each other soon."

The Pleiadian ship cloaked and returned toward Earth Orbit.

Earth and Pleiadian Command receive the message minutes later:

"First contact confirmed. Hostile. Prepare for war."

In the shadows of the Andromeda frontier, the Keltthran Empire vowed vengeance.

"Admiral Lumeris, question," Grant says. "Have we started a war?"

The admiral acknowledges her question with a nod as he speaks into his watch. "Possibly. The Keltthran are vengeful."

Admiral Lumeris addresses the group. "I need to excuse myself for a few minutes. Is there a private place I can make a call?"

"Yes," says Jack. "My office at the end of the hall on the right."

Muncie approaches Admiral Grant. "What can you tell me? Can you make sense of this?"

"Not right now, Colonel, but it was quite a meeting. Did you know about the Keltthran?" asked the admiral.

"Not really. We just kind of stepped in it when we created the Neuro Helix Protocol—accidentally acquired a new quantum organ in our prefrontal region. Created the equipment that facilitates communication with an expanded consciousness. It's a whole new experience. Our knowledge seems to expand with it."

"Admiral, I know this is not the time, and I don't mean to be blunt, but have we met before?"

Admiral Grant interrupts, "That's quite all right, Colonel. As I understand, you have a delightful habit of being blunt."

Muncie, taken aback, considers her next words carefully. "I see you have read my evaluation."

Sara smiles.

"I cannot help thinking we have met?" says Muncie.

Admiral Grant thinks for a minute then says quietly, "Colonel, I do not know, but let me think about it."

She turns and walks away, and then joins the two generals who were conversing.

"Okay, Generals, what's on your mind?"

"Well, we were expecting beer and pretzels," says General Wallace. They all chuckle.

Wallace continues, "What we've seen here is nothing like we have ever seen before. Tell us what you know, Sara."

"I can't—it's well above your current security level. But understand this: I have talked to my people and asked them to read both of you in. Frankly, I don't know what the outcome will be. If you do not make it, I advise signing a nondisclosure agreement and retiring, or just forget this ever happened. Soon, things are going to get dicey and dangerous. My feeling is the world will become a whole new battle theater, and so will the space around it."

Muncie thinks she looks haggard, like she had just lost a gunfight. Then he remembers *Eagles Nest Burning*. He looks for her, but she is with the other officers walking out the door. *Another time*, he thinks.

Muncie continues to observe the generals and Grant. The admiral had returned from Jack's office and began talking to Grant and the ambassadors. Muncie keeps thinking these people had met before.

After a short meeting with Grant and the female ambassador, the admiral announces that he wants to address the group.

After conferring with the ambassador and his ship, Admiral Grant continues, "Things are now in the works. It is my considered opinion that Sol is now a prime candidate for retaliation from the Keltthran Empire. Therefore, I have informed the Naval High Command on Vega Prime and requested reinforcements. There will be three more cruisers, three additional destroyers, and eighteen Fast Strike Interceptors. One Cruiser to replace our current one in orbit. This same defensive configuration has neutered a combined assembly of four hundred cruisers, and destroyers, and two dreadnoughts, and they didn't even break a sweat, as you would say on this planet. I will add this to form a new defensive armada. It will include ten complements of trained enhanced telepathic warriors. In the meantime, I have dispatched our destroyer to place added sensors near the Andromeda and Nexus Prime Gates, as well as in strategic locations in deep space. Our ETAs for the updated battle fleet are as follows: in seven days, the Interceptors along with the Destroyers will arrive; in eight days, one new Cruiser; and in eleven days, the last two Cruisers. At that time, we will depart for Vega Prime."

"Admiral," Muncie asks, "what is the probability that the Keltthran will retaliate?"

"Very probable," says the admiral.

"Admiral," Muncie presses, "what happens if there are more than four hundred vessels?"

"Then we will call for more capital ships and other assets, and we may take some casualties. Should they overwhelm us, Earth will take casualties," the admiral says solemnly. "Let me also say this: Currently, in high orbit in your solar system, is a battle-tested Destroyer and Cruiser. This isn't their first rodeo—whatever that is—and, as you say, there are four highly trained telepathic battle groups on the Cruiser. The Destroyer has two battle-confirmed telepathic crews. These weapons alone can collectively devastate an assault group of over four hundred assets. Any other questions?"

"Okay, Ambassador Lyara has some things to discuss, Ambassador," he continues.

"We are curious about the enhanced consciousness signatures. How did you do it, and can you show us?"

Jack answers affirmatively and motions to Lyara to follow him. They huddle up and begin talking.

Meanwhile, Admiral Grant had put on a pair of AirPods and is busy talking to her wristwatch. "Senator, they already know. We must read them in," speaks Grant. "Senator, the event that just took place in our solar system is only the beginning. The Pleiadeans have defeated a Keltthran Cruiser and two Destroyers at the edge of our solar system. According to the admiral, they will be back. We must prepare now and quietly make the most powerful people in our government aware of the risk. The cat is not out of the bag yet. When the Keltthran show up, what are we going to tell the President, the people?"

Grant waits patiently as the Senator finishes. Then she replies, "Senator, you and your group knew this would happen sometime. Sometime has arrived! The Generals will not forget this episode. They must be read in. We need more military support. There are now two generals and a retired air force colonel, who is about to be recalled to active duty, in the know. They will make great allies."

Bill and Muncie are both watching Grant. They give each other a glance. Then Muncie says, "Bill, I think the admiral knows something, and the two Generals don't."

"Agreed, but I'm concerned how this affects us," says Bill.

"We are now exposed to a deep secret that this government wants kept quiet. What do we do next, Bill? Look—Grant is talking to our Generals," Muncie says.

"All right, Generals," Grant states, smiling. "We have a security problem, gentlemen. You are not in possession of a high enough security clearance. I am recommending that it be remedied. So first, let me hear your questions or thoughts. I'm not saying I will answer them. So what are they?"

Captain Serenya with a phaser

The generals and Admiral Grant start a quiet, private discussion. Bill and Muncie observe, shaking their heads and conducting an animated discussion. Finally, it breaks up, the generals shaking their heads in agreement. Muncie thinks, *I wonder what agreement means. Are they in or out?*

"Ambassadors and Admirals," Jack announces, "let's go upstairs to Lab 3, and we will answer your questions and show you our equipment."

So off they go, leaving the military men. Bill and Muncie are left on the first floor, staring at each other. The admiral is busy talking to his wristwatch, quite absorbed, in the corner near the door, accompanied by a Pleiadian captain who had just arrived.

Bill and Muncie approach the Terran Officers and start talking.

Gen. Wallace asks, "Bill, what is your assessment of the equipment, the meeting, and the skirmish?"

"General, I'm overwhelmed. I can't believe my ears and eyes. I went through the Neuro Helix Protocol and my abilities are enhanced. I can telepathically communicate with Muncie anytime I want. That same thing goes for the people in my organization who underwent the protocol. My understanding of the equipment and how it operates is elevated. I enjoy reading—I can read three hundred pages in about forty minutes, and my comprehension is improved."

"So," says General Wallace, "what am I thinking right now?"

Bill responds, "Well, General, it's foggy, and I am new. Here's what I am copying: 'You're worried about Bobby and his current assignment… ummm…okay, you put that thought away, you are onto something new…I am losing it.'"

Muncie jumps in. "General McDougal, you are thinking about how beautiful Captain Serenya is…oops, it is gone now. What—whoa, now you are imagining what the Cruiser looks like. Admiral Grant, you are thinking about the same thing I am—*Eagles Nest Burning.* That is where we ran into each other."

"Hmm, Muncie, you are a pest," the admiral speaks. "But yes, it just came into my consciousness."

"Admiral Grant, I would really like to have a tour of the Cruiser," says Muncie.

"Let me see if I can sweet-talk the admiral." Off she goes to the conference corner with the captain and the admiral.

Grant asks, "Serenya, how have you been? Any good luck?"

Serenya replies, "Things have been active, especially in this part of the Milky Way. We may have a war brewing here and back on Eridani. Command is more upset about Eridani than here."

Grant asks, "So what's the problem on Eridani?"

Serenya says, "The Kovari Council and some very bad Thetas are working with them. One of the Thetas got self-elected as administrator of a city. She is ruining law enforcement and appointing her crooked cronies to leadership positions. Nothing except her agenda gets done. The Alphas are rebelling. You can just smell the Kovari's hand in this."

"Admiral Grant, we are going upstairs to join the ambassadors. Wanna come?" says Muncie.

"Yes, I'll be right there," replies the admiral.

Muncie waits for the admiral. When she breaks off conversation, it is just Muncie and the admiral walking together.

"I have been thinking about our discussion. Does the phrase 'Eagles Nest Burning' mean anything to you?" asks Muncie. She squinches up her eyes like she was thinking hard about something—or is pissed.

Muncie sees an aha moment in the admiral's eyes—he recognized that look. That look is exactly what he saw in the after-action meeting between the army, air force, and navy as a young lieutenant. She is pissed about a leak in security. Turns out, there was a mole in the hole. Nobody ever found out who it was.

Muncie says, "Aha, I recognize that look, Admiral. We have met. I will say—you have aged well. When we first met a few weeks ago, you looked like you had just lost a gunfight with the Galactic Empire. I don't see that look now." He says it humorously.

Admiral Grant stops, calmly looks him in the eyes, and with a smile like a predator getting ready to strike, says quietly and sweetly, "Oh no, Colonel, we WON that fight."

Muncie looks verklempt as he watches the admiral turn and calmly walk away. He thinks, *Oh shit, what have we here?*"

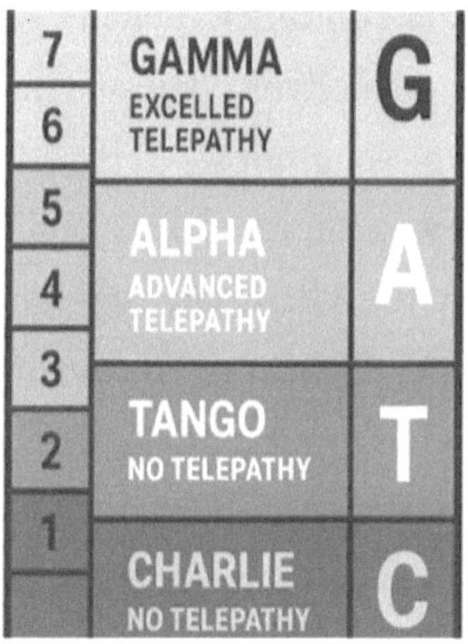

Working Consciousness Levels

When they arrive, Ambassador Lyara explains the Pleiadian consciousness. She says that Pleiadians, at the age of adolescence, are trained in raising their own consciousness. "The process is systematic, well documented, and taught by trained personnel. It adapts over time as we discover changes. There are seven layers of working consciousness, then you get into spirituality. At birth and in early adolescence, most are at level 1. Upon reaching what you call adolescence, the training begins. The goal is to raise that to level 3 and above. This takes about nine to ten months. We can measure these levels, as there are neurophysiological attributes we check before and after training. This process is identical for all Pleiadians, but Gammas and

Alphas each receive more specific training as they gravitate toward level 5. The reason is simple: 'Knowledge is structured in consciousness.'[7] The higher the consciousness, the more knowledge. Currently, Gammas can reach a top consciousness of 7, Alphas, about 5.5. We have a few Alphas reach 6—that is a relatively new phenomenon. Training is both general and specific for each classification [Alpha, Gamma]. Now, higher consciousness is not the end. It is the start."

Jack asks if she can explain more about their interpretation of consciousness.

"Yes," Lyara replies. "Consciousness is more of a frequency, with a corresponding amplitude or energy. In our training, Pleiadians learn to control and manipulate both. Do you know most computers can be manipulated by consciousness?"

Jack admits he does not.

"That brings me to the Neuro Helix," Lyara continues. "Thank you, Jack, for the earlier demonstration. We propose Genodyne's equipment be used to instantiate Neuro Helix into a select group—or groups—of Pleiadian volunteers. This takes place on our Cruiser, and we require working replicas of the Quantum Neural Frequency Enhancer, the Quantum Resonance Scanner, and the Quantum Resonance Frequency Generator for our own use, and a second replication for Genodyne, which we manufacture. You have shown us your schematics of the equipment. We want those schematics and the right to modify them to Pleiadian standards. Once finished, they remain in the possession of Genodyne and Pleiades. There are semiannual royalty payments made to Genodyne, and working replicated equipment given to Genodyne. Consider this part of the first cultural exchange and trade agreement."

Jack asks what kind of changes are planned.

"As a start," Lyara explains, "we replace your 'rare earth metals' with some of ours. They are not on your periodic table, yet. Having your schematics, I memorize them, and with your permission, I transmit them telepathically to our scientists and engineers on our Cruiser. Those Pleiadian modifications are all Genodyne property."

Jack then asks if they intend to sell replicas to customers.

"Yes," Lyara confirms. "We, of course, pay Genodyne additional annual royalties for each unit sold."

"What about future upgrades?" Jack asks. "Do you still pay royalties?"

"Yes," Lyara replies. "Our typical contracts run for one or two standard centuries. They are then renegotiated."

Jack asks who their customers are.

"Every ship in the entire Pleiadian Navy, and very possibly the entire Vega Navy," Lyara answers.

Jack comically asks if that is a lot of ships.

"Yes," both ambassadors and the admiral say, chuckling together.

Episode 5

New Minds for the Neuro Helix and the Aurora Cruiser Visit

Muncie approaches Grant and asks if she has spoken to the admiral.

"Yes, he wants to run it by the ambassador first. I think she is in charge," Grant replies.

They watch as Admiral Lumeris approaches the ambassador and starts talking. Casually, Muncie asks, "How do you know the admiral?"

Grant draws her head back, looks Muncie in the eyes, and says, "Muncie, you really are a pest."

Muncie interrupts, "Where have I heard that before?"

Grant groans and looks away, staring at the ongoing conversation between the admiral and Ambassador. After a short time, Admiral Lumeris looks at Grant. The Ambassador meets Grant's eyes and motions for her to approach.

Lyara asks, "Admiral, you want a tour, is that right?"

Grant replies, "Yes."

Lyara continues, "What about the two Generals? Can you vouch for them?"

Grant answers, "I'm working on that. I have worked with both, currently at a lower rank. They are good men, and I trust them. I have talked to my people, and they are considering reading them in. I have no idea if that will happen. I have advised the generals, if they are not admitted, they should sign an NDA and retire. There is a possibility that when all this gets published, they could return as consultants. We will need them."

"What about Muncie? Sara." asked the Ambassador. "Grant replies "I had worked with him a long time ago and have read his current file. He is a good man and officer. He will be recalled to active duty very soon and I have confidence in his abilities. He is one of the first Terrans to take the Neuro Helix protocol and is recognized as a HI Alpha Consciousness."

Lyara asks, "Admiral, what are you going to show them?"

Lumeris replies, "The cargo bay, bridge, Admiral's Quarters, Captain's Quarters, the mess hall, the food reproduction, and the rail guns and torpedoes. Then, high tea in the observation room."

Lyara nods. "Not too much detail on the weapons. Okay."

Admiral Grant walks back with Admiral Lumeris. "Okay," she says. "How do you want to do this, Admiral?"

Lumeris asks, "What are your schedules tomorrow? Everyone. So tomorrow, how about we pick you up here at 8 a.m.? We will have brunch on the bridge and wrap up around 1 p.m., returning a little after 3 p.m. Captain Serenya, we will have brunch in the admiral's Quarters tomorrow at 10:00 a.m. Will you join us?"

Serenya smiles and replies, "Of course, Admiral," then turns and walks away.

Sunday, January 11, 2026, 8:00 a.m. SOL time

Pleiadian Cruiser Bridge

As Jack and company wait for their 'ride' at Genodyne, it is a grey day in East Tennessee. Overcast shrouds the morning from Knoxville to Asheville. At precisely 8 a.m., the Pleiadian scout shuttle quietly decloaks in the parking lot. Jack looks at its size and figures it holds about a dozen people.

The door opens, and Captain Serenya steps out. "Good morning. It's a great day for a trip. Our Cruiser is parked on the other side of the moon, about forty-five minutes away after clearing Earth's atmosphere. We will depart after everyone is aboard for the short trip. The admiral wants me to send his apologies for not being here, as important activities require his attention. Your pilot today is Aer-lye, a fighter pilot who volunteered for this assignment. Should something happen to him, I am fully qualified to pilot this ship."

Once they are loaded and strapped in, the scout cloaks, then silently and smoothly lifts off. Through the windows, Earth fades as they enter the clouds, then break through the cloud cover and quickly leave Earth's atmosphere.

Serenya explains, "We select a specific geolocation to attain breakout and pass through the atmosphere. We adjust this every time to avoid current air traffic and possible collisions. Our return approach will be different. Coming and going presents unique challenges based on air traffic and weather."

As the group watches Earth grow smaller, the artificial gravitic system engages, and there is no sensation of zero gravity. Serenya explains, "We will not feel any zero gravity as we have an artificial gravitic system. The system creates a gravity of about 60 percent of Earth's. You may feel lighter but haven't lost any weight. Our Cruiser creates a gravity of about 65 percent of Earth's, so the transition from shuttle to Cruiser will be seamless."

The trip ends quickly as they approach the Aurora Military Airport. Landing is smooth in the now decloaked shuttle. After exiting, they gawk at the huge airport landing bay.

Aurora Class Cruiser: Air Landing Bay

The air landing bay of the Aurora Class Cruiser is a vast, high-ceilinged expanse designed to accommodate interstellar fleet operations. Visitors see polished metallic floors and illuminated guidance strips that direct incoming and outgoing craft. Panoramic observation windows offer glimpses of deep space, while reinforced blast doors ensure security during high-alert operations.

The bay buzzes with activity. Sleek fighter squadrons stand ready for rapid deployment, their silvery-blue hulls gleaming under ambient lighting. Robust military landing craft occupy the central staging area, designed to transport troops and equipment with modular interiors. Versatile shuttle craft, such as the Pleiadian scout, are stationed near the entrance for personnel transfers, diplomatic missions, and reconnaissance.

Advanced maintenance facilities surround the craft, including repair bays with robotic arms and quantum resonance tools. Automated refueling stations service multiple ships simultaneously. Secure storage units manage spare parts and munitions through a sophisticated inventory system. Adjacent crew lounges and briefing rooms provide spaces for debriefing, mission planning, and rest.

Glass-enclosed traffic control towers overlook the bay, serving as the nerve center for all landing, launch, and maintenance activities. Together, these crew areas foster camaraderie and readiness, making the Aurora Class Cruiser not only a formidable vessel but a welcoming home for its diverse crew.

Jack and Muncie look at each other and mimic, "WOW." Muncie thinks strategically and tactically, wondering how the Genodyne equipment will fit with current operations.

Bridge: Aurora Class Cruiser

The bridge is a marvel of advanced engineering, serving as the nerve center for ship operations. Panoramic windows stretch across the front and sides, offering commanding views of deep space, distant stars, and planetary bodies. Cool blue and silver lighting reflects off sleek metallic surfaces, creating a high-tech ambiance.

At the heart of the bridge sits the central captain's chair, elevated and surrounded by sophisticated workstations. Each station is manned by Pleiadian crew members, distinguished by their silvery-blue skin and elegant uniforms. Officers engage in navigation, tactical coordination, weapons, engines, communications, and ship systems management. Holographic displays and advanced control consoles provide real-time data, tactical overlays, and mission updates.

The bridge design emphasizes efficiency and collaboration. Multiple workstations allow seamless coordination between departments, with crew constantly monitoring ship status, plotting courses, and responding to external threats.

Admiral's Quarters: Aurora Class Cruiser

"Welcome to the Admiral's Quarters," Serenya says. "We have the finest chefs from our planet with us, and they have prepared both Earthly and Pleiadian brunch. Our chefs come from the best in Pleiadian and Vega Collective. While they have not trained on Earth, we have access to some of the finest recipes in all worlds. Some foods are synthesized from special replicators but cooked in a special galley on the Aurora. Our chefs have other military occupations, but when we host special guests, they get to cook, create, and enjoy their work."

She introduces four beaming chefs, grinning ear to ear with their food on display. The admiral walks in, apologizes for being late, and gives Serenya a special nod. She excuses herself to attend other duties but plans to return. She helps herself to some Pleiadian delicacies before leaving. The admiral suggests everyone dig in and starts serving himself; others follow his lead.

Jack, Muncie, Kathy, and the group are overwhelmed. Jack turns to Muncie. "Are we in Disneyland?"

"Yeah!" Muncie responds.

Kathy thinks, "I bet Jeff would love this."

The admiral's quarters include a spacious lounge and meeting area bathed in soft blue and silver lighting that echoes the bridge. Plush seating surrounds a central glass table for strategic briefings or informal gatherings. Holographic displays provide tactical overlays or serene views of deep space.

Adjacent is a compact yet refined library. Shelves of physical and digital volumes line the walls, allowing study or reflection. Three advanced workstations with sleek consoles, holographic interfaces, and secure communication links let the admiral coordinate fleet operations, review intelligence, and manage ship systems.

The kitchen features modular appliances, a food synthesizer, and a small dining nook. A private bathroom includes a rainfall shower, smart mirrors, and integrated storage. The bedroom offers a large bed facing a panoramic window, soft textiles, adjustable lighting, and personal touches, providing a restful sanctuary with privacy and security.

“This is a great brunch,” Kathy says, breaking her usual silence from the weekend.

“So,” Admiral Grant begins, “Dr. Redkey, I have been following your published research and am curious about where this might lead.”

Kathy’s fingers curl tightly around the edge of the table, knuckles white. She shoots a desperate look at Jack, her breath shallow. Jack watches her, heart pounding. He wants to reach out, to pull her close and promise safety, but all he does is nod, hoping she sees the strength in his eyes. Inside, he rages—against the kidnappers, against the uncertainty, against his own helplessness. For Kathy, he forces calm, silently urging her to keep going.

Her voice cracks. “I don’t know what’s happening to me. The Neuro Helix…it’s like something inside me shifts, and I can’t tell if it’s real or just in my head.” She swallows hard, fighting back tears. “What if it changes us in ways we can’t control? What if my son—” Her words falter, the thought too heavy. “I want to believe this is good, but I’m terrified. What if we’re opening doors we can’t close?” The room seems to close in around her, every face watching, waiting for answers she doesn’t have.

Ambassador Lyara leans in, eyes intense. “Kathy, you’re not alone. We’ve wrestled with these fears ourselves. The Neuro Helix is uncharted territory—even for us. But together, we can find answers.”

Kathy’s heart hammers. Memories of the kidnapping flash behind her eyes, cold and vivid. She presses her lips together, voice barely audible. “I want to help, I really do. But every night, I wake up sweating, remembering the hands that grabbed me, the faces I couldn’t see. My son—he’s all I have. The thought of losing him terrifies me. I try to act brave, but the fear…it never leaves. I’m scared that whatever we’re doing here, whatever the Neuro Helix is, it might put him in danger again.”

At the sound of her voice fading, Admiral Lumeris, Lyara, and Serenya rise, urgency etched in their faces.

"Let us help!" Lyara says, fierce and protective.

Jack clenches his jaw, fists tight at his sides, wishing he could shield Kathy from every threat.

Bill Roberts steps forward, his presence commanding. "No one touches Kathy or Jeff. Not while we're here," he says, steel in his voice. "Last time, we fought back. Next time, we'll be ready."

Jack exhales, tension easing just a fraction. Kathy looks around, support palpable—but the shadow of fear lingers, refusing to let go.

Admiral Lumeris speaks first. "Bill, Jack, and all of you, we need assurances of your safety so we can help. Let us show you. This will be in conjunction with your company, Bill. I think you will find working with us a pleasure and profitable. We will in no way interfere with your contractual obligations, but we can add features that will amaze you."

Bill looks at Jack. "I would like to see them."

Kathy then speaks. "Ambassador Lyara, you mentioned 'working consciousness' yesterday. Can you explain? For instance, are there levels above 7? What are they, and how do they work?"

Lyara answers, "Of course. I can explain some of them. There are at least three, possibly four or more consciousness levels above 8. At level 8—and we still have these officers in the Navy—one becomes more guided toward the spiritual. At eight, most officers and enlisted personnel request removal from active fighting roles and move to strategic positions. At nine and ten, they move into retirement and concentrate more on good and evil. They provide that kind of role and support. None of the nines and tens are active in the military, but they still provide spiritual and strategic guidance.

There may be higher levels, but I am not aware of them. Frequency of consciousness, which we can measure, is the determining factor."

Aurora Observation Deck

Sunday, January 12, 12:44 p.m.

"It appears to us," Admiral Lumeris says, "that we in Pleiades have some things to learn and new activities to address. We need to experiment with the Neuro Helix, and Earth needs a trained cadre of Terran battle telepathics to defend itself. We can do that together. If Earth can supply qualified personnel, we will train them. Pleiades needs trained Pleiadian/Terran Neuro Helix warriors. We are here today to plan and activate those plans. It is now a very apparent need to protect this planet and start this project. The Keltthran are coming."

Muncie and Jack watch Grant. She has been talking to her wrist for a full twenty minutes. She finishes and whispers to Ambassador Lyara. The Ambassador shrugs and nods.

Next up, Admiral Sara Grant.

"Currently, the United States has developed a Black Book space fleet with the help of the Pleiades government and Navy," Grant says. "It is small by any standard, but it is a start. The admiral and I have been discussing what a Navy would look like with a shipload of Neuro Helix warriors aboard. It is time to start working toward that goal. We need to manufacture one hundred Quantum Neural Frequency Enhancers in the next month, deploy them on ships and in training facilities. We need Pleiadian and Terran volunteers to upgrade their consciousness with the Neuro Helix. I understand that Pleiades will train both Terrans and Pleiadians."

Admiral Lumeris speaks next. "We have analyzed the schematics we received yesterday and are prepared to start manufacturing tomorrow. Dr. Dunkirk, we have items to go over as soon as you are ready. Would tomorrow be all right?"

"How about now?" Jack asks.

"Follow me, Jack," Admiral Lumeris says. "Muncie, Kathy, Bill, come with me."

"Now you guys," Admiral Grant says, looking directly at Generals Wallace and McDougall, "we have a few things to discuss. Follow me."

Mining the Entry Gates

As the meeting breaks up and the Terrans return home, Admiral Lumeris signals Captain Serenya and the ambassadors.

"I am notifying Pleiades, Vega, and the Interstellar Council that Pleiades will be mining the Entry Gates to SOL. Serenya, when you return, see me. I have important work for you and Captain Theissan. Ambassadors, here is my reasoning: According to Galactic Law, mines can be used if a threat of planetary extinction or enslavement exists or can be proven. Given the history of the Keltthran, I am concerned Earth's population will be enslaved or eliminated and the planet plundered for minerals. Let me read you something."

"By All Rights"

The admiral begins. "Under the 'By All Rights' clause of Intergalactic Law, our protection claim is not only justified—it is unassailable. Ambassadors, I am informing you that I have contacted Pleiades Council of my intention to file a notification to mine the Entry Gates at SOL. Just an hour ago, I received permission to file this claim with Intergalactic Legal. Frankly, Pleiades turned this around faster than I could imagine. I am also asking for Diplomatic support."

"Well, Admiral," Ambassador Lyara says with a smile, "you have really stepped in it this time. Here is what you do not know: about a month ago, I wrote a similar request to Pleiades recommending we guard the gates. I did not mention mining specifically, but I hinted about it. Your memo dovetailed remarkably well with mine, which is why it turned around so quickly. We are on the same page. Next time, talk to me first. You know how easy I am to convince."

The admiral smiles. "Would you be willing to write the message to the Intergalactic Council? I will sign it."

"I already have. Check your email," the ambassador replies.

Admiral Grant and the Generals

"Generals," Admiral Grant begins, "you will be read into the project. Consider this your initial briefing. You will report to me and no one else. This project is above top secret, and your security clearances are being adjusted as we speak. We are extending our stay here to fourteen days. During this time, we are to quietly help Admiral Lumeris and the ambassadors. All three of us will undergo the Neuro Helix Protocol."

"Admiral, a question," General McDougall asks. "What if we do not want to undergo the procedure? Is there a medical reason for this?"

"No," Admiral Grant replies. "Then you will be dismissed. Is it your intention to refuse the protocol?"

The General thinks for a moment, then says, "No."

Admiral Grant acknowledges McDougall's answer and continues. "General Wallace, is it your intention to refuse the protocol?"

"No," Wallace answers.

Admiral Grant says, "Good. Tomorrow we start the protocol."

Episode 6

Thunder on Eridani

Genodyne Offices

Monday, January 12, 2026, 7:30 a.m.

Kathy, Jack, Muncie, Admiral Grant, and Generals Wallace and McDougal arrive and proceed to LAB 3.

"Oh, Kathy, did you mention the injection to the officers?" Jack asks, speaking within earshot.

Grant, Wallace, and McDougal exchange a glance and stare at Jack with what can only be described as an evil eye. Jack looks back and notices Kathy turning red, while Muncie chokes back a laugh. Jack laughs and says, pleasantly, "Coffee, anyone?"

The joke lands, and everyone starts laughing. "Well, that woke everyone up," Jack explains.

He continues, "The process itself takes about fifty minutes each. First, have you taken the IQ test last night so you're ready for the protocol? Good. Step one: hook you up to the scanner, looking for frequencies to measure and identify. Step two: take the protocol. Step three: break for coffee or hook back up to the scanner. Step four: take another IQ test, and Bob's your uncle. Okay, who's first?"

Sara Grant genuinely likes Jack. He's a breath of fresh air compared to all the military types she's spent her life with. She saunters up to him, looks him in the eyes, and smiles suggestively. "I'm going to hurt you," she says quietly.

Jack leans in and whispers, "Oh, please, don't let me get my hopes up."

They both crack up laughing. Grant quietly says to Jack, "I'm going to tell Muncie today."

Jack replies, "Make sure I'm there. I want to start the conversation." Grant nods.

"I have never seen Sara behave this way. Have you?" General Wallace remarks.

"No, I haven't. She's enjoying Jack's company," General McDougal replies.

As they all complete their protocol, General Wallace approaches Grant. "Admiral, what's next?"

"Next, we transfer to Jack's property and wait for the shuttle," she replies. "We'll be using his property for future transfers—it's more private and secure. We'll spend the next seven or eight days on the Cruiser and Destroyer observing. You two and Muncie will meet Admiral O'Brian on the Atlas, the Terran Destroyer involved in the EOS Gate skirmish. You'll also meet Lt. Commander Maelis, the Engineering Officer; Lt. Commander Ser-ly, the Weapons Officer; and Commander Ra-li, XO for the Cruiser. You may spend time on Destroyer D229 with Captain Theissan, who was also at EOS Gate. That's what's next, Generals. Oh, and Muncie, you'll be with the generals the rest of the time. If you need to run back to Earth each day, that's fine. We're going to upgrade your military resume. Generals and Muncie, you'll work closely with Captain Serenya; she will introduce you to key Pleiadian officers and train you on critical assets and protocols. Pay attention."

She takes Muncie aside and quietly says, "Your role at Genodyne will be modified. Jack and I will fill you in."

"I'll be tending to Admiral Lumeris and Ambassador Lyara. They're putting together a quick class on Interstellar Law just for us. Attendance is mandatory. Jack, you and Kathy are not required to attend. By the way, we'll be working twelve- to fourteen-hour days up there. Time is relative—it goes fast. Oh, I almost forgot: Jack and Kathy will spend some time with me today. The Pleiadians want more information on the equipment and the Neuro Helix. When the fourteen days are finished, we should have two sets of working equipment and sixteen Pleiadian volunteers trained with the Neuro Helix."

On the Bridge of the Terran Destroyer Atlas

General Wallace enters the bridge, where Admiral O'Brian reviews tactical logs.

General Wallace begins: "Admiral, good afternoon. I'm General Douglas Wallace. My colleagues are General William McDougal and retired Colonel Muncie Mathews. I've been wanting to ask about the EOS Gate skirmish. How did the crew handle themselves? What did you think when the Pleiadian Destroyer arrived?"

Admiral O'Brian responds, "Retired Colonel Mathews, please tell me how you fit into this motley group. I'm curious."

Muncie states, "Well, I was on track for General, but my wife died of cancer rather suddenly. My adolescent children were taking it badly. I had a choice: I couldn't abandon them. I knew the path to General would take me away for months as I padded my resume with all the right assignments. So I retired. I ended up at the TVA briefly, then faced the 2020 layoff. I made a deal with TVA to perform work on contract as long as I could pick my team. I knew the owner of a company called Genodyne and made a deal to rent space in his office building. Jack and I go back to undergrad; we reconnected later in life."

Admiral O'Brian responds, "So the EOS Gate…I remember my communications officer reporting something large coming through. We were training for this scenario. When the Pleiadian Destroyer D229 decloaked and challenged the Keltthran fleet, everything changed. Our crew had spent weeks mastering every inch of this bridge—coordinating navigation, tactical, and engineering until it was second nature. When the enemy appeared at EOS Gate, everyone knew their role. Shields were ready, weapons targeted. When D229 revealed itself, the Keltthran formation faltered. They refocused their weapons on D229, while we targeted the Keltthran Cruiser. In that instant, I thought the tide had turned—not just due to technology, but because of courage. The presence of the Pleiadian ship signaled we were not alone; together, we could push back even the fiercest assault. Frankly, I believe we would have prevailed, though we might have taken damage."

"General, did you hear what the Pleiadian Captain said to the Keltthran Admiral?"

Wallace says, "No."

"He said, 'Give my regards to the High Imperator,'" O'Brian reports.

The bridge cracks up laughing.

General Wallace says, "I've noticed Terrans adapt quickly. How steep is the learning curve for new recruits?"

"On the second-generation Atlas, it's challenging but manageable. The bridge layout is intuitive, and the tactical station integrates all weapons controls. Third-generation ships are easier—same layout with upgraded shields and some new tech, like quantum torpedoes. Training modules simulate real combat, so they're ready on duty," Admiral O'Brian admits.

"And weapons—what's your favorite?" Wallace inquires.

O'Brian responds, "I like the railguns—precision and raw power. But quantum torpedoes on new ships are impressive, bypassing most defenses."

Wallace says, "Sounds well equipped. One last thing—how's the food?"

O'Brian laughs. "Better than expected for a warship. The galley staff takes pride in their work. Fresh supplies arrive every few weeks, and chefs keep morale high. Terran stew after a long shift is unmatched. We do have automated printing, but the full staff prepares two fully cooked meals a day—usually breakfast and early dinner with all the trimmings."

"I'll hold you to that, Admiral."

Monday, January 12, 2026

Cargo Bay, Pleiadian Cruiser

Jack, Admiral Grant, and Kathy arrive on the Pleiadian Cruiser, depart the shuttle, and are met by Captain Serenya.

"Greetings. I hope your flight was comfortable. We're going to Engineering Lab 2. Two prototypes are ready to test. Preliminary tests ran into trouble. Jack, we could use your advice with the Pleiadian model."

Engineering Lab 2, Pleiadian Cruiser

Serenya: "Jack, here's the data from the Pleiadian model."

Jack: "Let's look at the diagnostics." He hands some to Kathy. "What did you notice during the last test?"

Kathy: "The neural interface keeps losing sync. Every time we ramp up the frequency, the feedback loop destabilizes. It's like the system can't lock onto the Pleiadian baseline."

Jack reviews the data. "Hmm. The prime and secondary harmonics are too noisy. The Pleiadian neural signature is more complex than we thought. Did anyone isolate the third harmonic?"

Kathy: "No, they focused on primary and secondary. You think the third harmonic is key?"

Jack: "Let's see if the third harmonic matches the conscious frequency the Pleiadians use for deep focus. If we set that as primary, the interface should stabilize."

Kathy adjusts the controls: "Switching to third harmonic... recalibrating...so far, so good."

Serenya: "The readings look strong. Frequency lock achieved. Jack, nice work!"

Jack: "You're welcome, Captain."

Kathy: "Remind me to buy you coffee back at Genodyne."

Jack: "Only fresh—no instant. Can we move into Terran and Pleiadian trials today?"

Serenya: "We have Pleiadian volunteers ready. A few Terran volunteers exist, about twenty-five aboard. I'm unsure why more haven't signed up; I'll investigate."

Two hours later

Pleiades Cruiser, Alcyone Prime, Bridge Conference Room

Jack and Kathy peek into the conference room. Captain Serenya looks up from her reading.

Serenya: "Come in, sit down. Apparently, Terrans didn't know they could volunteer; we've corrected that. We now have fleet wide, eighty-three Terran volunteers. I want you to meet two from Destroyer D229—Merope 2—who participated in the EOS Gate skirmish. Come in, Sergeant Aer-tyer and Private Mike Cochran. Dr. Jack Dunkirk, Dr. Kathy Redkey, please sit. Admiral Grant should hear this too. You're not in trouble, but they should hear what you did at EOS Gate."

Admiral Grant enters. "Be seated." Jack grins.

Grant smiles back before focusing on Serenya.

Serenya: "These two soldiers are part of the new Pleiades-Terran Telepathic Battle Group. They participated in a successful psychic attack on the Keltthran Cruiser that entered SOL at EOS Gate. Sergeant, please continue."

Sergeant Aer-tyer: "I've served fifteen of your years. At the Gate, our officer of the day informed us we'd execute a Class 3 attack on the Keltthran. We organized quickly; as the captain engaged the Keltthran Admiral verbally, we infiltrated their control systems. Our main focus was the Cruiser, but we gained control of two other Destroyers as well. We worked on their weapons when they fired their first shot. We executed a micro jump, and when we recovered, neutralized their weapons and shields. I targeted the admiral while my young apprentice isolated him. I coached him stepwise to neutralize the target; by accident, a code made him sick."

Private Cochran: "If I had my way, he'd have had a bowel movement." Dead serious, he says.

Everyone laughs. Jack and Admiral Grant laugh hardest. Serenya chuckles and then resumes business.

Serenya: "These are our first two volunteers. Training starts in thirty-five minutes. It's not painful—expect greater mental and physical performance, higher IQ, and enhanced consciousness. We measure before and after. One standard hour per person. Grab a drink or light food and prepare. Indeed, they begin their trek to higher consciousness. After two days, 83 new Terrans and 216 new Pleiadians have a Neuro Helix. Training continues as they learn to use the Neuro Helix with their new consciousness level."

February 16, 2026

Pleiadian Destroyer, Merope 2

Captain Serenya was doing TDY on the Pleiadian Destroyer alongside Colonel Muncie Mathews when LT Aer-ryr informed her, "Captain, you're wanted in the wardroom off the Captain's Quarters."

When Serenya arrived, Mathews and Captain Theissan were deep in conversation. Muncie turned toward her and said, relieved to see her, "I'm telling you, Captain, I'm concerned. The Terrans are still learning at a rapid rate. I don't know how they'll perform in this kind of combat scenario."

Serenya frowned. "What scenario is that?"

"There's an uprising—and possible slaughter—going on in EOS 3, on the second planet in Epsilon Eridani,[8] located in the constellation Eridanus. It's 10.52 light-years from Earth," Muncie explained. "We are the closest unit, and the fleet thought we were the logical choice. Our mission is to rescue and remove all Pleiadians and any populations threatened on the planet. The former population was around 7,600 three years ago. Intel is sketchy, but the Thetas are committing genocide against the Alphas and Gammas. This attack is complicated. The Thetas are using a psycho-telepathic assault, supported by about five hundred troops of unknown origin with energized weapons. Our main purpose is to rescue the remaining Pleiadians and, if possible, capture one or more of the unknown troops."

Serenya grimaced. "Ugh. I don't like this at all. How many Marines are on the Merope 2?"

"144," Theissan replied.

Muncie asked, "And on the Thunder Lance?"

"About two hundred," Theissan answered.

Muncie did a quick calculation. "So that's 344 Marines against 500 unknowns. How many Marines on the Merope 2 have the Neuro Helix?"

"Eighteen," Theissan said.

"And on the Thunder Lance?" Muncie pressed.

"I don't know," Theissan admitted.

"What happens if there's a dreadnought hovering over the planet?" Serenya asked.

"We'll need the Cruiser," Theissan replied.

Serenya continued, "What are the numbers for the other two Cruisers—the Galactic Synthesis and Singularity Crest?"

General Wallace, quietly observing from the entrance, asked casually, "Can we defeat the Thetas with the Neuro Helix?"

Muncie examined his data pad. "We need to upgrade the entire fleet with the Neuro Helix, especially those going on this mission. Get me Admiral Grant and Jack Dunkirk."

Everyone paused, staring at one another, absorbing the weight of the statement. Muncie added, "What do the Pleiades know about the Thetas' psycho-telepathic abilities?"

Serenya answered, "Very little. We also need intelligence on Kovari military capabilities."

Two hours later, the group assembled in the Bridge Wardroom of Crusader Prime. Present were Admiral Ter-Ir-Tye, Grant, Captains Theissan and Serenya, Colonel Mathews, Generals Wallace and McDougal, and—via hologram—Jack Dunkirk and Kathy Redkey.

Muncie addressed the assembled officers. "Thank you all for meeting on such short notice. We are evaluating options and trying to get our minds around the rescue operation on Epsilon Eridani. We're dealing with some unknowns: the naval strength and makeup of the Kovari military, whether they have a dreadnought, how many Thetas are active, and how many Marines are involved. We suspect the Thetas and Kovari Council are supporting this to expand their territory. At 10.5 light-years and a jump line of four to five days, Sol could be in their gunsights very quickly, assuming their jump engines work like ours. Time is critical—the attack is ongoing."

He continued, addressing the group. "Admirals and everyone, here is our proposed battle plan: Rescue at Epsilon Eridani."

The plan began with the deployment of **Crusader Prime**, volunteered by the admiral, while mines at the EOS Gate were placed on standby for safe passage. Stealth drones were sent through the gate to gather intelligence, allowing the team to adapt the plan in real time.

The initial approach was in stealth mode, with an **All-Neuro Helix attack force aboard Crusader Prime**, supported by a telepathic group tasked with disrupting enemy communications and attempting to disable enemy ship systems and Theta forces.

The second Cruiser and Destroyer followed in stealth, analyzing sensor data and adapting as needed. The Destroyer provided immediate support to Crusader Prime via communications interference and sensor spoofing.

A **phased attack profile** was proposed: after Crusader Prime jumped, the second Cruiser and Destroyer would enter the jump gate two days later, ready to support Crusader Prime or remain in stealth. If a dreadnought were present, the second Cruiser's role would be to distract it long enough for the rescue to succeed.

Muncie emphasized, "The plan's effectiveness relies on timely intelligence and the performance of new technology, with the ultimate goal of rescuing threatened populations and capturing unknown enemy troops, despite the urgency and uncertainties surrounding the operation. We'll be attacking an unknown number of Thetas, supported by a large force of Kovari Marines. Time is of the essence. The plan is robust, but it hinges on intelligence quality and the performance of our new technology."

Probability Table

Scenario	Probability of Success
All enhancements and intelligence work	70–85%
Intelligence incomplete, enhancements work	60–75%
Enhancements or distraction fail	<40%

Conclusion

With the addition of stealth drone intelligence gathering, the battle plan's probability of success improves to 70–85 percent, provided the Neuro Helix enhancements and distraction tactics were effective. If either factor failed, the probability would drop sharply. The plan is robust, but it depends heavily on the quality of intelligence and the performance of the new technology.

Admiral Ter-Ir-Tye is the first to speak. "Those are not very good outcomes. Is there anything we can do to improve them?"

Serenya replies carefully, "Yes, if we had better intel and more time."

Ter-Ir-Tye's tone is flat.

"Well, Generals, your thoughts?" the admiral prompts.

General Wallace leans back slightly. "Going up against a dreadnought…I don't like the odds."

General McDougal nods in agreement. "The urgency factor is overwhelming."

"Thank you all for your thoughts. Captain Serenya, your opinion, please," the admiral says, turning to her.

Serenya speaks carefully, her body language betraying her discomfort. "I'm torn between the urgency and the probability. We don't have a very good outlook. Basically, we are rolling the dice, as I've come to understand from talking with the Terrans. But…I say we go."

Muncie interjects, firm. "The captain has expressed exactly my thoughts. We go!"

Ter-Ir-Tye looks at Wallace and McDougal. "Go or no-go?"

Reluctantly, they answer in near unison: "Go."

"Okay. Start preparations. Theissan, your Destroyer will be part of the second wave. Serenya, do you have a preference for the other Cruiser?"

"The Iron Phalanx, sir," she says.

"Muncie, you and Serenya will assist and lead the ground forces. Any questions?" the admiral asks.

"No," they reply simultaneously.

Muncie glances at Serenya and begins to wonder if he had the stamina for an excursion like this.

Serenya catches his eye, grabs his arm, and says with a hint of a smile, "Come on, Colonel. We're going to the shooting range."

Episode 7

Eridani Rescue

The next few hours are a hectic blur as crews assemble, ships are readied for Jump and Battle protocols, and a new order comes down. All crew members in the SOL fleet receive the Neuro Helix Protocol. During this same timeframe, Colonel Muncie familiarizes himself with the standard armament of a Pleiadian Marine—the Gamma Hand Phaser and Rifle.

The Genodyne Quantum Armament is partially installed on the Crusader Prime and the Iron Phalanx. A full complement would be thirty-eight units for a Cruiser and seventeen units for a Destroyer, enhancing the performance of the Telepathic Battle Group. With the new Jump and Quantum drives, it is possible to speed up the trip by adjusting the Jump Drives to the resonance frequency of the Jump Line, using less energy and reducing travel time. The return trip might require a different resonant frequency, which must be recalculated. This theoretical approach will be tested with the rescue project. Another test is planned: what happens if the Neuro Helix of the combined Telepathic Battle Groups operates at the resonance frequency of the Jump Line?

The SOL fleet assembles swiftly at a designated point, with all ships undergoing rigorous armament checks. Marines equip themselves with Gamma Hand Phasers and Rifles, while Genodyne Quantum Armament installations are prioritized for the Crusader Prime and Iron Phalanx. The Telepathic Battle Group receives briefings on new operational protocols, emphasizing collective defense against Gamma consciousness attacks from the Thetas.

A critical component of the mission is the administration of the Neuro Helix Protocol, mandatory for all personnel to enhance neural resilience against psychic attacks. Only 66 percent of the crew completes the protocol at launch, prompting emergency medical teams to accelerate administration, prioritizing frontline marines and command staff. Engineering teams simultaneously calibrate the Jump Drives to the resonance frequency of the Jump Line, optimizing energy efficiency and preparing contingency plans for recalibration on the return journey.

As the fleet prepares for launch, intelligence estimates that five hundred Thetas and an unknown number of marine forces oppose the rescue, prompting tactical overlays and simulations stressing coordinated action by the Telepathic Battle Groups. All personnel are briefed on the threat posed by the Thetas' Gamma consciousness and the necessity of the Neuro Helix Protocol. A real-time intelligence review is scheduled at the end of the Jump Line, using data from scout drones in Epsilon Eridani. With a fleet-wide alert issued, last-minute checks ensure armament, drive calibration, and protocol compliance as the fleet braces for the unknown, united by the hope of rescuing seven thousand souls.

Muncie notices General Wallace on board and salutes. The general returns the salute and walks over. "Greetings, General. Coming with us to the big game?"

"Muncie," the general says, "this is a tough one. You've just processed back into the Air Force, and now you're going into combat. Stay safe, keep your head on a swivel, my friend."

Muncie gestures toward a nearby marine. "General, I want you to meet someone. First Sergeant," he calls out. The sergeant approaches and salutes. Muncie and the general return it.

"General, I would like to introduce First Sergeant Stanley Adams. He's twenty-five, a Terran Marine, and has been with the Black Marine detachment for almost two years. He's a Gamma and has a Silver Star. I'll be in his capable hands." Muncie adds, "The Pleiadians believe their O5s and O6s need combat field experience. It boosts morale. Sergeant, at ease. Tell the general a little about your experience."

The sergeant briefly describes seven firefights and the types of skirmishes he's involved in, noting that this mission appears the most dangerous. General Wallace thanks him and turns to Muncie. "Be careful, Muncie."

East Tennessee, Genodyne Offices

Jack Dunkirk addresses the assembled employees. "Everyone, please sit down. I know there have been rumors, so let me put them to rest. Muncie Mathews has been recalled to active duty. For Genodyne Engineering, Bradley Yorktown is promoted to manager of the dual position Muncie held. This is no surprise, as Muncie's duties are increasingly dedicated to Genodyne Research. His position in Research will remain open for now. A few personnel updates: we have a new CFO starting Monday, Molly Pillsbury, and Annette Thornberry is promoted to Controller. Congratulations, Annette. Genodyne Engineering had a banner year—revenues are up over 50 percent, and we have potential new clients. Genodyne Research has a new client; I'm 90 percent confident this will be a long-term arrangement. They are a very private company and will remain so. The future looks profitable and exciting. We are also planning new offices, including a private airport. Any questions?"

Cruiser, Crusader Prime, heading toward EOS Jump Gate

Muncie turns to Serenya. "Here we go. How many expeditions like this have you been on?"

Serenya considers. "I don't know exactly…about ten, I'd guess. I will say this one is the most dangerous. We don't have good intel, and that concerns me. On the others, good intel saves lives."

"It worries me too," Muncie admits. "Good luck."

As they approach the Jump Gate exit at Epsilon Eridani, an announcement echoes through the ship: "All hands, Jump Gate exit in thirty standard minutes. Battle groups, assemble outside your assigned shuttles. Prepare for Jump."

Serenya briefs Muncie. "Here's how it's going to work. Once through the gate, we receive intel from the drones. We analyze it quickly and put together our initial plan, including the locations of the Pleiadians and other personnel. I need your help with that. The drones provide an AI-generated suggestion for Plan A. We must have a Plan B no matter what. With the time crunch, we will need to mobilize and execute immediately. I'm not comfortable with this, but lives are on the line, including ours. The procedure is to remain stealthy: get in, get out, rescue the souls, and never be seen. The drones also give a timeline of what has occurred on the ground, and if possible, intercept communications from the Kovari ship. There are too many unknowns, but we do have a Cruiser and Destroyer due to arrive in four days."

Muncie nods grimly, salutes the captain, and returns to his group. Early planning places Captain Serenya as the lead, with Muncie as the ready reserve. The first group consists of nine telepaths and a mixed Pleiadian/Terran Marine battle group of sixty soldiers.

An announcement follows: "All hands to battle stations. We are exiting the EOS Gate. Rescuers, board your jump ships. Officers, you should have the intel from the drones."

"Affirmative, Rescue One," Serenya confirms.

"Affirmative, Rescue Two," Muncie responds.

"Affirmative, Rescue Prime," answers the Crusader Prime command officer.

Telepathic groups discuss the land layout and rescue planning. Three groups require extraction: the largest, about 2,500, is trapped and engaged in combat; a second group, roughly 1,100, is retreating but harassed by hostiles; a third, about fifty, is stationary and out of danger. Three shuttle teams are assigned: Rescue Three, a group of twenty-five Marines and three telepaths, extracts the fifty souls easiest to rescue; Rescue One, fifty Marines and five telepaths; Rescue Two, forty Marines and four telepaths. The Marine contingent is divided: one-third Terrans using kinetic weapons, the rest using Gamma rifles. The Cruiser executes a micro jump after each shuttle release, announcing their positions despite stealth. The Crusader Prime houses up to fifteen fighter groups; ten are on ready alert.

"Rescue Prime, this is Rescue One, ready for launch."

"Rescue One, standby. We are not in position."

"Rescue One, affirmative, standing by," Muncie confirms.

As Muncie prepares for descent, his mind whirls with tension and responsibility. The operation is fraught with uncertainty; intel is incomplete, and hostiles are unknown. Every decision could mean life or death.

He reflects on operational protocols, aware that collective defense relies on the Telepathic Battle Groups and the Neuro Helix Protocol—only 68 percent of personnel have completed it at launch. Emergency medical teams race to close the gap, but the risk of psychic attacks from the Thetas looms.

As the shuttle nears the planet, Muncie exchanges telepathic thoughts with Serenya. She admits this is the most dangerous mission they have ever faced due to incomplete intel, heightening his sense of responsibility. The plan: remain stealthy, execute the rescue swiftly, and avoid detection. He knows adaptability is key. The fate of seven thousand souls weighs on his shoulders.

Aboard the Crusader Prime, Captain Aer-Rye orders, "Flight Commander, prepare three combat flights for active duty."

"Commander, Air Group," the communications officer responds. "Commander Lumrys, copy, Captain."

Aboard the Iron Phalanx and Merope 2, Captain Aev-Ti, General McDougal, Captain Theissan, and Admiral Grant listen on private communications. Grant says, "He's getting ready to launch fighters. What does he know?"

Captain Theissan replies, "Something's not right. It must be spooking him. He would launch only if he feels the ground troops are in over their heads or to defend the Crusader Prime."

Then the comms light up: "Rescue Prime, this is Rescue Three. We are landing one kilometer from target."

"Rescue Prime, this is Rescue Two. We are taking fire."

"Rescue Prime, this is Rescue One. Landed, moving toward Position One."

"Rescue Prime, this is Rescue Three. We're pinned down. This is a trap. Need air support."

“Affirmative. Air support en route,” comes the reply.

Commander Air Group orders, “Launch all three fighter groups. Flight One to Rescue Three, Flight Two to support Rescue Two, Flight Three to support Rescue One.”

With the rescue underway and heavy resistance, horror and fear spread across the generals and admiral. The captains are engaged in the attack, mentally taking notes. This is what awaits them upon arrival.

Captain Aer-Ry reports, “Telepathic Battle Group, status report.”

Telepathic Battle Group Report: Dreadnaught Engagement

Lead Telepath (via secure comms)

“Command, this is Telepathic Battle Group reporting on the dreadnaught attack. We engage the hostiles. Current status: 39% of enemy forces are unconscious, 42% confirmed dead. Hostiles actively attack our personnel. Approximately 22% remain inactive; these do not possess the Neuro Helix and are not participating in the assault. Our success rate here is significantly higher than on the ground, where resistance remains strong and our telepathic tactics are less effective. We recommend that fighter groups initiate a psycho-telepathic attack to suppress hostile ground forces and improve our chances of rescue. Please advise on authorization for this tactic.”

Command responds, “Telepathic Battle Group, your report is received. Fighter command is reviewing your recommendation. Stand by for further instructions. Continue to attack and monitor enemy activity. Maintain defensive protocols and update us on any changes in hostile behavior.”

Over the comms, Rescue Two calls, “Rescue Prime, this is Rescue Two. Grey Fox is down. Send reinforcements.”

"Rescue Prime, this is Rescue One. Blue Angel is down," Rescue One reports.

"Rescue Prime, this is Rescue Three. Where the hell is air support? We're being overrun."

Rescue Prime replies, "Rescue Two, reinforcements are leaving. ETA fourteen standard minutes."

Lieutenant Byer-ty observes from Fighter One Squad. "We're over target—we see you, Rescue Three. Keep your heads down. We're coming in for a close shave." Fighters scream down in a dive, leveling out for their run. The overpowered Gamma Cannon flares, spewing Gamma plasma with overcharged nano munitions. One round lands fifty meters to the left. Thump. Thump. Thump. To the right, ninety meters away, a massive explosion erupts, followed by another, then silence.

A surrender emerges—white flags rise. Eleven, no—fifteen—twenty-two souls step out of the forest line, walking toward the clearing.

"Rescue Three, this is Fighter One. Your status?"

"Fighter One, we're okay. Continue support. We accept prisoners, load the dead and wounded, and return to the ship. We require an escort."

The corporal oversees loading the prisoners, the wounded, and the dead. The 'prisoners' claim they were captives; among them, nine are Terrans, the rest Pleiadian. Two fighters escort the shuttle back to the ship. The remaining two break off to join the hunt.

The Emotional Impact on the Command Team

As battle chatter echoes through the private communications radio aboard the Iron Phalanx and Merope 2, Admiral Grant, General McDougal, and Captain Theissan feel a wave of shock and dread. The announcement that Blue Angel and Grey Fox are both down hits them hard, sending a chill through the command team. The loss of these two key leaders during the rescue operation underscores the gravity of the situation and signals that the mission may be on the verge of disaster.

Admiral Grant and General McDougal's faces reflect horror and fear. Losing such vital members amid the operation forces them to confront the possibility of mission failure and the risk to the rescue teams. Their thoughts race with the implications, especially considering the thousands of lives hanging in the balance. The tension in the room is unmistakable, each officer acutely aware that the outcome of the battle now depends on their decisions and the resilience of their teams.

Captain Theissan listens closely to the ongoing chatter, gripped by anxiety and a profound sense of responsibility. Knowing their colleagues are in imminent danger sharpens his focus, fueling his determination to do everything possible to support the rescue efforts. He mentally documents the unfolding events, preparing for the critical moments when his choices could mean the difference between survival and defeat.

Despite the fear and uncertainty, the command team finds unity in their urgent resolve. The burden of leadership in combat becomes clear: constant uncertainty, heavy responsibility, and the courage required to continue pressing forward even when hope seems faint. Bound together by commitment to the mission and to each other, they face the unfolding crisis with unwavering determination.

Admiral Grant addresses the group telepathically. "We can't wait two days. Captains, how soon can we jump?"

Captain Theissan answers, "Eight to ten standard hours."

Captain Aev-Ti responds, "We are not that close. We have a lot of people needing the Neuro Helix protocol."

Admiral Grant presses, "Captain, if we skip the testing and just apply the protocol, that saves about forty minutes per person. If we do four at a time, that's twelve an hour. How many people do you have?"

"About one hundred and twenty," the captain says.

"We could complete that during the jump. Do you have the new jump drives?" Grant asks.

"Yes," the captain replies.

"Okay, I'm authorizing the abbreviation of Neuro Helix testing as an emergency protocol for the support of the Eridani Rescue. I need to contact Admiral Lumeris. Start loading."

One hour later

Admiral Grant speaks over secure channels. "Admiral Lumeris, thank you for taking my call. Are you aware of the current situation on Eridani and the rescue?"

"I am aware of the rescue effort but not the current situation," replies Lumeris.

"Okay, here's what I need. I'm signing an authorization to facilitate and speed up the Neuro Helix Protocol for the Iron Phalanx and Merope 2. No testing before or after—we just apply the protocol and move on. Intel from the rescue operation shows the protocol is saving lives. The Neuro Helix prevents telepathic attacks from causing damage. More data will be in the after-action report. Next, I need authorization to immediately launch the second wave jump to Eridani. Based on current intel, more support is needed for existing rescue efforts," Grant explains.

"Now the bad news: Blue Angel and Grey Fox are reported as down. They are casualties—no intel on whether they are dead. Each led a rescue team that was ambushed and now fights without leadership. The situation is fluid. With the new jump drives, we can arrive in two and a half days. We need to move quickly, Admiral," Grant pleads.

Admiral Lumeris sighs. "I see the problem and your solution. You need a temporary order to speed up the Neuro Helix protocol—this is saving lives. Second, an order to immediately launch the second wave to Eridani as additional support."

"Yes," Grant confirms.

"Okay, continue your process. I'll create the orders. Let me know the status of Blue Angel and Grey Fox as soon as possible."

"Understood," Grant replies.

Iron Phalanx

As the words "Blue Angel down, Grey Fox down" echo through the private communications radio, General McDougal feels a cold wave of shock and dread. These aren't just call signs—they are colleagues, leaders, and friends whose presence has anchored countless missions. The sudden loss strikes at the heart of the command team, forcing him to confront the brutal reality of combat: even the most capable can fall when the stakes are highest.

His mind races with memories of shared briefings, tense operations, and moments of camaraderie. Knowing that both Blue Angel and Grey Fox are casualties—without confirmation of their fate—leaves him grappling with uncertainty and fear. He understands, perhaps more than anyone, the ripple effect this will have on the rescue teams now fighting without their leaders. The mission, already fraught with risk, teeters on the edge of disaster.

Yet amid the horror and anxiety, General McDougal feels the heavy mantle of responsibility settle on his shoulders. The fate of thousands depends on the resilience and unity of the command teams.

In that moment, he is reminded of the true cost of leadership in battle: constant uncertainty, the courage to continue when hope seems faint, and the unwavering commitment to the mission and to those who serve alongside him. It has been a long time since he has been in battle, and those emotions return with a force he has not felt in years.

Episode 8

The Rescue Has Consequences

Merope 2 Destroyer D229 Capt. Theissan, Second Wave

"Captain," calls Adm. Grant, "we're approved. I'm just notified, and Captain Aev-Ti, General McDougal—Captain Theissan, are we ready to jump?"

"Yes, we're one hop away from the Gate," replies Captain Theissan. "You need to notify Captain Aev-Ti. We need to jump now."

"Already done, Admiral."

Five minutes later, they reach the EOS Gate and, as a group, jump to Eridani toward an uncertain rescue operation. ETA is two and a half days.

Crusader Prime, Captain Aer-Rye

"Commander Air Group, prepare four groups for launch with air-to-air weapons," barks Captain Aer-Rye. "Command, do not charge the rail guns, but target the dreadnaught."

Telepathic Battle Group 1: "Captain, the dreadnaught is getting ready to launch fighters."

Battle Scene: The Rescue at Eridani

The Crusader Prime hovers high above the embattled plains of Eridani. Though its stealth shields are engaged, the Kovari dreadnaught has scored some lucky shots, leaving its hull scarred but defiant.

On the ground, chaos reigns. The Pleiades marines—two-thirds Pleiadian, one-third Terran—press forward under heavy fire. The Terrans, clutching their kinetic rifles, move with grim determination. Each shot tears through the stealthy Kovari battle suits, revealing the enemy where others see only empty air.

The blue blocker sunglasses worn by three Terran marines with Rescue 1 give them a crucial edge, allowing them to spot and target the invisible Kovari marines. Around them, the ground is littered with casualties—both Pleiadian and Terran.

The Gamma rifles wielded by the Pleiades marines flash, but their beams don't always kill or knock the enemy down. The Kovari armor is superior, leaving the Terrans' kinetic rounds as the most reliable means of stopping them.

Realizing the tide is turning, Kovari officers bark new orders. Their weapons shift focus, targeting the Terrans with a barrage of energy bolts. Yet instead of killing outright, many of the shots knock the Terrans unconscious or leave them wounded. The Pleiades marines, less fortunate, suffer painful wounds but hold their ground, rallying behind the Terrans' effectiveness.

Above, the roar of engines signals the arrival of Pleiades fighter aircraft, launched earlier to support the beleaguered ground troops. Their strafing runs shatter Kovari positions and give the marines a much-needed reprieve.

Sensing the shift, the Kovari captain aboard the dreadnaught orders a squadron of fighters to intercept. In response, the Pleiades scramble additional planes, weaving through the sky to protect their comrades below.

Inside the Kovari dreadnaught, tension mounts. The Lt. Commander, voice sharp, admonishes the remaining crew as a telepathic assault from the Pleiades disables much of their flight crew and marines. Only the Thetas—a few elite Kovari and Pleiades Gamma soldiers—remain active, their resolve tested as the battle intensifies.

Despite mounting casualties, the rescue teams press their advantage. The Terrans' kinetic weapons continue to punch through Kovari lines, and the combined air and ground assault slowly shifts momentum. The Kovari fight back with everything they have, but the resilience of the rescue teams—and the edge provided by Terran technology and Pleiadian telepathy—begins to show.

The Lull in Battle

The thunder of engines fades as the Pleiades fighters pull back, leaving a tense silence over the battered plains. Amid the wreckage, the rescue teams seize the moment to tend to their wounded.

Rescue 1: Blue Angel's Aid

Behind a shattered transport, Blue Angel slumps against the hull, blood seeping from a wound in her arm. Her mind drifts.

I don't feel that much pain. Am I dying?

Corporal Lin: "Stay with me, Captain. The air cover's gone for now, and we've got a window to patch you up and get you out of here."

Blue Angel (thinking): *I can still see them—Kovari marines moving in the haze. Those blue blockers are a lifesaver. What is that humming? It feels... comforting.*

Corporal Lin: "You spotted three before anyone else. Without you, Rescue 1 would've walked right into their line."

Blue Angel (thinking): *Terran rounds—one shot, and they're down. That's what's keeping us alive. The Neuro Helix...it's humming, steady, like it's keeping me here.*

Corporal Lin: "We'll get you back in the fight. Just hold on."

Rescue 2: Grey Fox's Aid

Across the battlefield, Grey Fox lies in the shadow of rubble, breathing shallowly. Sgt. Adams works quickly, applying pressure to a wound.

Sgt. Adams: "Colonel, you're with Rescue 2. You took a hit, but you're not out. Stay with me. We're working on you—you're going to make it!"

Grey Fox (weakly): "Saw their officer...he was aiming for the Terrans. Lucky for us, their tech isn't always lethal."

Sgt. Adams: "Colonel, some of us have been killed. The Kovari changed something—upped the power. They're hitting harder now. We held the line, but it's getting worse."

Grey Fox: "More Terran rifles...more blue blockers. We're winning—but it's close. Sergeant, set the Gamma rifles to high."

Grey Fox begins to lose consciousness. The Neuro Helix hums softly, a low, comforting vibration, like a distant purr. He slips under.

Sgt. Adams: "They're set on low—switching to high now! All units, Gamma rifles to high power!"

He glances skyward as the distant rumble of engines signals fighters regrouping.

Bridge of the Crusader Prime

The bridge is alive with tension as alarms echo and tactical displays flicker.

Captain Aer-Rye: "Commander, report. What are the Kovari doing?"

Commander: "Sir, the dreadnaught is launching fighters—vectoring straight for our support flights."

Captain Aer-Rye: "We can't lose air cover. Launch four fighter squads. Intercept and neutralize any Kovari fighters targeting ground support."

Tactical Officer: "Aye, Captain. Four squads launching now."

Comms Officer: "Captain, message from ground command: Terran kinetic weapons are far more effective than Gamma rifles. They're requesting reinforcements equipped with kinetics."

Captain Aer-Rye: "Weapons, inventory check. How many kinetic rifles remain?"

Weapons Officer: "One hundred total, sir. Thirty-three deployed. Sixty-seven remain."

Captain Aer-Rye: "Good. I need sixty-seven volunteers—Terrans preferred. Issue each a kinetic rifle and prepare for immediate deployment."

Captain Aer-Rye (over intercom): "All personnel: if you possess blue blocker sunglasses, turn them in for reissue to ground forces. Volunteers may retain theirs. Report to the armory."

Minutes later…

Logistics Officer: "Captain, we have twenty-one blue blockers. All are with volunteers."

Captain Aer-Rye: "Excellent. Eight to Rescue 1, seven to Rescue 2. Ensure all recipients are briefed—they're critical for spotting stealth units."

Dogfight over Eridani

Delta Five: "Missiles incoming! Ground support can't intercept—Echo Two, break off!"

Echo Two: "Negative, Delta Five! We're pinned—need air cover!"

Squadron Leader: "Air support, engage at range! Target missile platforms!"

Delta Seven: "Two Kovari down—missile lock successful!"

Delta Four: "Nine Kovari fighters down! Ground support, you're clear!"

Wingman: "Thirteen of us made it back, sir."

Captain Aer-Rye: "You did more than enough. Prepare for rearm. This fight isn't over."

On the ground

Ground Commander (Comms): "Crusader Prime, we have thirty-eight casualties. Two shuttles destroyed. We're down to thirty-three operational—fourteen at Rescue 2, nineteen at Rescue 1. Request immediate escort."

Shuttle Pilot: "Shuttle Alpha lifting with wounded—need cover!"

Fighter Escort: "Alpha, Delta Four here. We're on your wing."

The Stalemate at Eridani

Rescue 1 Marine: "Gamma rifles on high! It's working—keep pressure!"

Rescue 2 Volunteer: "Kinetic rounds—one shot, one wound. We're making progress!"

Above, telepathic attacks intensify, disrupting Kovari coordination, weapons, and shields. This appears to be limiting the dreadnaught's effectiveness.

Despite gains, the field remains littered with casualties. Neither side breaks—the battle settles into a stalemate.

Medic: "Blue Angel's vitals are holding. Grey Fox is responding. They're fighters—they'll pull through."

Back aboard the Merope 2

Admiral Grant (Comms): "Admiral Lumeris, this is Grant. Volunteers made a difference. Gamma rifles on high and kinetic weapons are giving us an edge. Telepathic attacks are working—but casualties are high, and the Kovari are holding. Blue Angel and Grey Fox are critical but stable. At present, it's a stalemate. We are a day and a half out from Eridani."

Admiral Lumeris (Reply): "Understood, Grant. We need a new strategy—and you are it. Make it work."

Admiral Grant: "Copy that."

Scene: The Arrival of Iron Phalanx and Merope 2

Plan A

The battered plains of Eridani tremble as the sky splits open with the arrival of the Iron Phalanx and Merope 2. The Crusader Prime hangs in high orbit, its shields flickering, weapons sparking from the last Kovari barrage. On the ground, medics work frantically among the wounded, while exhausted marines cling to makeshift barricades, eyes fixed on the horizon for any sign of hope.

Suddenly, the comms crackle to life. "This is Admiral Grant of the Merope 2. Iron Phalanx, you are clear to deploy. All units, prepare for coordinated assault," Admiral Grant announces.

From the Iron Phalanx, a formation of heavy assault walkers—each bristling with kinetic cannons and shield projectors—descends in drop pods, slamming into the no-man's-land between the Pleiades and Kovari lines. Their arrival sends shockwaves through the battlefield, scattering debris and forcing both sides to take cover.

But the real surprise comes from above. The Merope 2, using its reserves of power, initiates a planetary-scale EMP pulse—targeted to disrupt only Kovari technology, thanks to intelligence gathered from captured enemy gear. The pulse ripples across the field, shorting out Kovari stealth fields, disabling their advanced targeting systems, and leaving their dreadnaught momentarily blind.

Seizing the moment, the Iron Phalanx walkers advance, their shield projectors forming a mobile barrier that allows medics and engineers to retrieve the wounded and reinforce the front lines. Behind them, Pleiades and Terran marines surge forward, kinetic rifles blazing, blue blocker sunglasses glinting in the harsh Eridani sun.

On the bridge of the Crusader Prime, Captain Aer-Rye watches as the tide shifts. "All remaining fighters, launch now. Target Kovari positions while their systems are down. Ground teams, advance under Iron Phalanx cover," he orders.

The Kovari, caught off guard by the EMP and the sudden armored assault, falter. Their lines buckle as the combined might of the Iron Phalanx, Merope 2's support, and the rejuvenated ground forces press the attack. For the first time in hours, hope surges through the allied ranks.

As the dust settles, the battered but unbroken survivors realize the battle is not yet won—but with the Iron Phalanx and Merope 2 in the fight, victory is finally within reach.

Kovari Counterattack

The Shadow Reversal

The battlefield's silence after the EMP pulse shatters as a chilling broadcast echoes from the Kovari dreadnaught. The Kovari Lt. Commander, voice cold and resolute, issues orders across encrypted channels. "All units, initiate Protocol Shadow Reversal. Activate reserve stealth fields. Prepare for coordinated counterstrike."

Within moments, the battered Kovari lines surge with renewed purpose. Engineers scramble to deploy portable shield generators—smaller, but resistant to EMP effects. These devices create pockets of cover, allowing

Kovari marines to regroup and tend to their wounded. The Thetas, elite Kovari Gamma soldiers, rally at the front, their armor flickering as backup stealth modules come online.

Above, the Kovari dreadnaught, still partially blind, launches a final wave of fighters—each equipped with analog targeting systems immune to EMP disruption. These fighters fly erratic patterns, using the chaos below to mask their approach. Their objective: disrupt the Iron Phalanx walkers and sever the link between allied ground and air forces.

On the ground, Kovari officers bark new orders. "Deploy smoke and holographic decoys. Target the Iron Phalanx shield projectors—overload them with concentrated fire," one officer commands.

Kovari marines, using their superior mobility, dart between cover, launching coordinated barrages at the advancing walkers. The air shimmers with energy as holographic illusions create phantom squads, sowing confusion among the Pleiades and Terran defenders.

Meanwhile, Kovari telepaths—long held in reserve—unleash a psychic barrage aimed at disrupting allied communications and sowing doubt in the minds of the exhausted marines. The battlefield becomes a maelstrom of deception, misdirection, and psychological warfare.

Despite their losses, the Kovari counterattack is relentless. Their tactics shift from brute force to cunning: sabotage, infiltration, and psychological pressure. The Iron Phalanx walkers, though formidable, find themselves isolated as Kovari strike teams target their shield generators and supply lines. To break the isolation, the walkers regroup and form a moving safe area of protection.

As the sun dips below the horizon, the battle for Eridani enters a new phase—one defined not by overwhelming firepower, but by shadows, illusions, and the indomitable will of both sides.

Allied Commanders' Response

Turning the Tide Again

The Kovari counterattack sweeps across the plains of Eridani like a shadow reborn. EMP-resistant shield generators flicker to life, stealth modules reactivate, and analog-guided fighters scream overhead. Holographic decoys and smoke sow chaos among the allied lines, while psychic assaults batter the minds of weary marines.

On the battered bridge of the Crusader Prime, Captain Aer-Rye's eyes dart across the tactical display. "They're adapting—fast. Commander, status on Iron Phalanx?" he asks.

"Shields holding, but the Kovari are targeting projectors. We're seeing infiltration attempts—saboteurs in the rear lines," the commander replies.

Admiral Grant's voice cuts in over the comms from Merope 2. "We anticipated a counterstrike. Activate Protocol Mirror: deploy counter-psychic teams and signal all units to switch to analog comms. Iron Phalanx, initiate shield rotation pattern Delta—do not let them focus fire on any one projector," she instructs.

On the ground, field commanders relay updated orders. "All squads: ignore visual contacts unless confirmed by thermal or kinetic signature. Decoys are everywhere. Medics, keep the wounded under cover—Kovari infiltrators are targeting triage zones."

Aboard the Iron Phalanx walkers, engineers reroute shield power, cycling projectors in unpredictable patterns. "If they can't predict our shield coverage, they can't concentrate their fire," one engineer mutters.

Meanwhile, allied telepaths—long held in reserve—link minds to form a psychic bulwark, blunting the Kovari's mental assault. "Hold the line.

Focus on unity and comms disruption. Their illusions cannot break our will," their leader intones.

Above, the battered remnants of the Pleiades air wing coordinate with ground spotters using coded light signals. "Target confirmed—analog fighters at grid seven. All interceptors, converge and engage," a squadron leader directs.

As the Kovari press their advantage, the allied response remains swift, adaptive, and unyielding. The battle becomes a contest not just of firepower, but of wits, resilience, and the ability to improvise under fire.

Ground Troops' Experience

Victory on Eridani

The battered ground teams huddle behind shattered barricades, the air thick with smoke and the scent of scorched earth. The Kovari counterattack unleashes chaos—holographic decoys flicker in the haze, psychic assaults batter exhausted minds, and EMP-resistant shield generators create shifting pockets of enemy cover.

Orders crackle over analog comms. "Ignore visual contacts unless confirmed by thermal or kinetic signature. Decoys are everywhere. Medics, keep the wounded under cover—Kovari infiltrators are targeting triage zones," comes the directive.

Engineers aboard Iron Phalanx walkers cycle shield projectors in unpredictable patterns, exploiting openings and attacking in irregular bursts, denying the Kovari a clear target.

In the trenches, marines grip their kinetic rifles, blue blocker sunglasses glinting as they scan for real threats. The psychic bulwark formed by allied telepaths restores clarity and unity. “Hold the line. Their illusions cannot break our will,” the telepathic leader reinforces.

“Gamma rifles on high. It’s working—keep the pressure up,” one Rescue 1 marine shouts.

“Kinetic rounds—one shot, one wound. We’re making progress,” a Rescue 2 volunteer adds.

Gradually, the tide turns. The Kovari’s illusions falter under strict verification protocols, and their infiltrators are isolated and neutralized. Coordinated air-ground operations, using nondigital signals, break the enemy’s erratic fighter tactics. The combined firepower of kinetic rifles and Gamma beams begins to breach EMP-resistant shields, forcing the Kovari to withdraw.

Rescue Summary and Repatriation Plans

Of the 3,600 souls trapped on Eridani, only 2,347 survive the siege. Among them, three hundred and fifty are Terran, while the remainder are Pleiadian. The survivors—many wounded and traumatized—are stabilized in medical tents and aboard allied ships.

Amid the devastation, hope emerges in two remarkable recoveries: Grey Fox and Blue Angel, both previously listed as critical, now show strong signs of recovery under close medical supervision. The medical teams observe something extraordinary—the two appear to have healed themselves.

When brought aboard the Crusader Prime, both are in critical condition. Examination reveals that Grey Fox’s internal injuries have expelled shrapnel, relocating it beneath the skin for removal. Scar tissue has already formed where the organs have healed. Blue Angel sustains similar injuries, which have also remarkably repaired themselves.

It is noted that 90 percent of the wounded who exhibit accelerated healing possess a Neuro Helix.

Plans for repatriation are already underway. The majority of Pleiadian survivors will be transported back to their home worlds aboard the Merope 2 and supporting vessels, accompanied by medical teams and counselors to ensure recovery and reintegration. Terran survivors will be prepared for return to Earth via secured shuttles with escort fighters, coordinated by diplomatic liaisons for debriefing and reintegration.

The operation prioritizes care for all survivors, honors the fallen, and ensures a dignified return to their families and nations. The remarkable recoveries of Grey Fox and Blue Angel become symbols of resilience. As Kovari forces collect their wounded, they withdraw to safe ground and return to their ship.

The battle for Eridani is over—but its consequences are only beginning.

Episode 9

Eridani Paradox

Command Conference Room

Orbiting Eridani

The command conference room thrums with tension, the hum of starship systems barely masking the anxiety etched on every officer's face. Admiral Grant stands apart, her gaze locked on the holographic map of Eridani—a planet battered by decades of betrayal and war. As the flickering light dances across the faces of Captain Aer-Rye, Captain Aevaris, and Captain Aev-Ti, the stakes crystallize: their next decision determines not just survival, but the soul of Eridani itself. Now, with the Kovari dreadnaught looming above, its presence a silent threat, Admiral Grant knows the fate of Eridani hangs by a thread. The officers debate, tension mounting.

"If Eridani is worth keeping, we'll know soon enough," Admiral Grant declares, her voice grave. "But if the data shows it's not viable, we pull out—no hesitation, no more losses."

The silence that follows is heavy, each officer contemplating the fate of their people. The decision crystallizes: launch the satellite, gather hard mineral data, and—crucially—interview the survivors. Their stories, shaped by decades of occupation, betrayal, and resilience, will guide the next steps.

Outside, Eridani glows—a world battered, but not yet broken, its future poised on the edge, its fate not yet decided.

The officers nod, the decision made. Outside the viewport, Eridani glows—a world on the brink.

Situation Report: Battle Aftermath

The current battleground is secure and quiet. Both settlements are badly damaged. Agricultural production is impaired, and light mining in both areas remains partially damaged but repairable and functional. Kovari and Theta presence on the planet is significantly diminished.

The Kovari dreadnaught is largely inert due to the success of the psycho-telepathic attack by the Pleiadians. This attack renders a large population of Kovari marine and naval personnel unconscious or "inert." Sixteen prisoners are detained aboard the Cruiser *Iron Phalanx*. The ground belongs to the Pleiadian and Terran soldiers.

Situation Report: Officers' Impressions

The officers have a protective stance toward their people. The Pleiadians, having won the battle, remain deeply concerned about any group—especially recent arrivals—imposing servitude or causing unrest. The officers prioritize justice, reconciliation, and the restoration of autonomy for all affected, with special attention to those injured by the Thetas' actions.

"These actions—the corruption of the vote, the packing of the bureaucracy, the interruption of regular supplies—have caused deep anger," Captain Aevaris says.

"This demands a cautious approach," Captain Aev-Ti adds. "Healing and unity must come before any mining operations—or we evacuate the planet entirely."

Situation Report: New Facts

Interviews with Pleiadian Population and Historical Records

"What was shared—and previously unknown—is this," an intelligence officer reports. "When the first Kovari arrive, they number about twelve thousand souls. About three thousand are Thetas. One thousand Thetas and five thousand Kovari are transferred to the colony near the minor mineral area. The existing Pleiadian colonies at the time have populations of about 4,500 each."

"That amounts to an invasion," Captain Aer-Rye says.

"Both Pleiadian populations are outnumbered after living on the planet for one hundred and thirty-two years," the officer continues. "These are quiet, peaceful settlements. What happens next is frightening. Regular shipments of food and supplies from Vega stop. The Pleiadian population begins to starve. Meanwhile, Kovari supply ships arrive regularly."

"They ask for help?" Admiral Grant asks.

"Yes. The Pleiadians ask for help, but the Kovari and Thetas refuse. Communications become irregular. That's when the riots start. This appears to be a planned event of extermination—and someone in Vega is involved."

A heavy silence settles over the room.

Admiral Lumeris is livid. *We have a spy. Who would do this?*

Admiral Grant stands in shock. The captains are already shifting toward retaliation.

"We declare war," one of them says. "We execute all Kovari and Thetas."

Admiral Lumeris raises a hand, then turns and leaves the room. Moments later, in his quarters, he opens a secure channel.

"Ambassador Lyara," he says, his tone controlled but urgent. "We have a traitor in our midst. We need to meet."

Strategic Deliberation

The officers, deeply aware of Eridani's mineral wealth and its long-standing Pleiadian population, confront the scars left by invasion, supply disruption, and orchestrated starvation. Their sense of moral responsibility sharpens under the weight of the Kovari and Theta actions.

"Under Galactic Law—the 'By All Rights' clause—Eridani belongs to the Pleiades," Captain Aer-Rye states.

"And we enforce that," Captain Aev-Ti replies.

While Eridani has not been formally declared a protectorate, the officers' stance and the settlement's history have effectively created one. Their priorities—safety, autonomy, and well-being—align with that framework.

"We do not abandon the planet," Admiral Grant says firmly. "One hundred thirty-two years of settlement—we don't walk away from that."

"Restarting the population and removing the Kovari and Thetas gives a clean slate," Captain Aevaris says, "but it risks retaliation."

"Negotiation?" another officer suggests.

“Shared mineral rights could foster peace,” Aev-Ti says. “A council, environmental protections—structured cooperation.”

“Reparations?”

“Uncollectable,” Admiral Grant answers. “And enforcement would be…problematic.”

The room falls into measured silence as each option is weighed.

Conclusion

Ultimately, the consensus forms.

“We remove the Kovari and the Thetas,” Admiral Grant states. “Not out of vengeance—but under law, and for protection.”

No one objects.

Eridani’s future will be shaped by Pleiadian autonomy, justice, and unity—with no place for those who orchestrated its suffering.

The discussion stops.

“We speak to the Kovari dreadnaught’s captain,” Admiral Grant says. “Now.”

Communications Link

Pleiadian Command Ship to Kovari Dreadnaught

The bridge of the Pleiadian command ship is tense. Officers gather around as Cap. Aev-Ti opens a channel to the Kovari Dreadnaught. The holo-display flickers, revealing Iear-II, a Theta in Kovari uniform.

Capt. Aev-Ti: "This is Captain Aev-Ti of the Pleiadian fleet. I wish to speak with your captain."

Lt. Commander Iear-II (Theta): "The captain is unavailable. I am currently handling communications."

Capt. Aev-Ti (narrowing eyes, voice measured): "You're a Theta. What is a Pleiadian doing on a Kovari ship?"

Lt. Commander Iear-II (expression hardening): "I'm not Pleiadian anymore. I am Kovari."

A ripple of surprise moves through the Pleiadian and Terran officers on the bridge. Some exchange incredulous glances.

Capt. Aev-Ti: "That's...unexpected. Why aren't you moving? Your ship's been stationary for quite some time."

Lt. Commander Iear-II: "Our engines won't start."

Aev-Ti glances at the telepathic officers, who suppress smiles. Aev-Ti gives a subtle signal, and the telepathics ease their interference.

Capt. Aev-Ti: "Perhaps that will change now. Try your engines again."

On the display, Iear-II checks her console. The distant rumble of engines is heard. The Kovari dreadnaught begins to move.

Capt. Aev-Ti: "Why do you call yourself Kovari? Did you grow up among them?"

Lt. Commander Iear-II: "No. My family moved there. We chose a new allegiance."

The officers absorb this revelation, recognizing the complexity of identity and loyalty in the aftermath of conflict.

Capt. Aev-Ti: "Be advised: The Kovari are in violation of Interstellar Law and will be reported. However, if you wish to open diplomatic relations, that door remains open. The choice is yours. One more thing—we have sixteen prisoners. When do you want to collect them?"

Lt. Commander Iear-II: "I will relay your message to the captain. Keep the prisoners."

The channel closes. The Pleiadian and Terran officers exchange thoughtful looks, aware that the next steps will shape the fate of Eridani and its people.

Iron Phalanx Conference Room

Holographic Link Active

The officers gather around the table. Holographic images of Captain Aer-Rye and General Wallace flicker to life. On the Merope 2, Captain Theissan and Admiral Grant join the call.

Capt. Aev-Ti (leaning back, smirking): "Well, that was a conversation for the history books. I ask for the Kovari captain, and I get a Theta with a new accent and attitude."

Gen. McDougal (raising an eyebrow): "Did she really say, 'I'm not Pleiadian anymore, I'm Kovari?' That's one way to switch teams. I hope she kept the jersey."

Capt. Aer-Rye (hologram flickering, voice dry): "Let's just hope she didn't take our playbook with her. The last thing we need is Kovari running Pleiadian tactics. I find her reply regarding prisoners to be extremely cold-hearted."

Gen. Wallace (chuckling): "Maybe she just wants better cafeteria food. I hear the Kovari have excellent rations—if you like everything gray and slightly crunchy."

Capt. Theissan (from Merope 2, serious): "Jokes aside, her loyalty shift is troubling. If Thetas can just 'become' Kovari, what does that mean for our intelligence? For our security?"

Adm. Grant (nodding, then grinning): "It means we need to start screening for accent changes and sudden cravings for Kovari cuisine. But seriously, it's a reminder—identity is fluid in times of war."

Capt. Aev-Ti (smiling): "I admit, I enjoy the part where their engines don't start. The look on her face when I tell her to try again—priceless. Telepathics deserve a medal for subtle sabotage."

Gen. McDougal (mock stern): "Next time, let's make their coffee machine break down. That'll really test their loyalty."

Capt. Aer-Rye (returning to business): "Humor aside, the Kovari are in violation of Interstellar Law. We make that clear. But I leave the door open for diplomacy. If they want to talk, we'll listen—after they fix their engines and their attitude."

Adm. Grant (thoughtful): "Diplomacy is good. But let's keep our telepathics on standby. Just in case they try to 'switch teams' again."

Capt. Theissan (smiling): "Agreed. And maybe next time, we'll ask for a Kovari captain and get a Terran chef. At least then we'll eat well."

The officers share a brief laugh, but the seriousness returns as they prepare for the next steps—balancing humor with the gravity of their responsibilities.

Iron Phalanx Conference Room

Strategic Planning

The officers remain gathered, the holographic link humming quietly. The mood shifts from banter to focused determination.

The officers lean in, the tension palpable as the fate of Eridani hung in the balance. Captain Aev-Ti's voice cut through the hum of the conference room. "We've shown the Kovari we're in control, but telepathic tricks won't hold forever. It's time for decisive action."

Gen. McDougal's fingers danced across the datapad, determination apparent in their eyes. "Our people come first. Reinforce the settlements, monitor Kovari movements—if they so much as twitch, we're ready."

Capt. Aer-Rye's hologram flickered, a wry smile breaking the seriousness. "If Thetas can swap loyalties so easily, maybe we should check ourselves for sudden cravings for gray, crunchy food."

Gen. Wallace's grin faded as resolve took over. "Jokes aside, we need to listen to our survivors. No more Kovari on Eridani. We owe it to those who've endured."

Captain Theissan's voice echoed from Merope 2, carrying the weight of the survivors' wishes. "The poll is clear: Most want to leave, but some hope to rebuild. If we stay, it must be by consensus—no Kovari allowed."

Admiral Grant's gaze lingered on the battered planet below. "A sanctuary, untouched by mining or military, could offer healing. But after all we've seen, I'm inclined to agree: Eridani needs a fresh start, free from Kovari influence."

Capt. Aev-Ti (smiling): "Council, sanctuary, and open dialogue. I too am seriously considering No More Kovari. Having thought about it, I can see both Mining and Refining on Eridani. That way the product can be delivered in prefinal status, and the buyers can turn it into whatever they want."

Gen. McDougal (dryly): "And if they don't, we sabotage their coffee machines next time. Frankly, I like the idea of no more Kovari."

Laughter ripples through the group, then fades.

Capt. Aer-Rye (standing, decisive): "Let's move. I'll handle ground coordination. Theissan, Grant—draft guidelines for rebuilding. McDougal, Wallace—tighten security and reach out to survivors. And someone check the mess hall for gray food."

General Wallace (quietly, cutting through the room): "How do we protect both Sol and Eridani?"

Silence follows, the enormity of the question settling over the group. The XO quietly contacted Admiral Lumeris and Ambassador Lyara, as the officers braced themselves for the challenges ahead.

Command Conference Room

Orbiting Eridani, Protection of Both Systems

The group gathers around the holographic display, the tension palpable as the fate of two worlds hangs in the balance.

As the officers gather in the Iron Phalanx Conference Room, the mood shifts from light banter to determined resolve. The team quickly agrees that telepathic tricks and diplomatic gestures are no longer enough. The new plan is clear: Eridani would be rebuilt with the help of the Pleiades, focusing on restoring homes and farms first. Mining and refining operations would be established at a safe distance from agricultural zones, ensuring the planet's recovery and sustainability.

In the quiet after the vote, Adm. Grant lingers at the viewport, watching Eridani's bruised landscape shimmers beneath the stars.

"We've chosen justice," Admiral Grant murmurs, her voice barely audible. "But justice comes at a cost."

Capt. Aev-Ti joins, their usual bravado subdued. "I keep thinking about the families—what they've lost, what we owe them. No more Kovari, no more councils. Just us—and the burden of rebuilding."

For a moment, the officers are not commanders; they're witnesses to the aftermath of war, united by the ache of responsibility and the hope that their choices would bring peace.

As security teams sweep through the battered settlements, the officers pause to listen—*really listen*—to the survivors. Faces gaunt from months of deprivation, voices trembling with both hope and anger, they recount the horrors of occupation: families torn apart, supplies withheld, and the constant threat of betrayal.

Gen. Wallace kneels beside a young Pleiadian, offering reassurance as tears streak the child's dust-covered cheeks.

"You're safe now," Wallace promises, the weight of responsibility heavy in their voice.

The decision to send prisoners to the Pleiades and separate the Kovari from the Thetas is not just protocol; it is a pledge to protect those who had endured and to begin the slow work of healing.

Admiral Grant lingers at the viewport long after protocol requires, her reflection faint against the glass as Eridani turns slowly below. The planet's surface is a mosaic of ruin—vast scars etched into continents; cities reduced to silent geometries of ash and shadow. From orbit, it is almost beautiful, the way a broken stone catches the starlight. Almost.

She knows better.

Eridani is not merely wounded. It is a warning.

Grant folds her hands behind her back and steadies herself as unease tightens in her chest. She had commanded fleets, ordered evacuations, watched lives hinge on her decisions—and yet this quiet moment unnerves her more than any battle. The devastation below whispers of failures that had once seemed impossible: Betrayals that began small. Leaders who hesitated. Supplies that ran out. Trust that fractured when it was needed most.

What if Earth is no different?

The question circles relentlessly. Earth, with its confidence, its arguments, its faith in distance and deterrence—is it truly prepared for what Eridani had faced? Could its systems endure sustained orbital bombardment? Could its people remain united when food dwindled, when fear outpaced reason, when survival demanded sacrifice instead of rhetoric?

Grant swallows.

She imagines the early days here, the assurances, the delays, the belief that diplomacy or technology would save them at the last moment. She had heard those same reassurances before. On Eridani, they had sounded just as convincing.

Her thoughts darken. If an invasion came, it wouldn't announce itself with clarity. It would arrive through confusion, misinformation, and hesitation. Through leaders afraid of being wrong. Through voices dismissed as alarmist until the sky itself fell.

She rests her palm against the viewport, as if the cold glass could anchor her resolve. Should she speak now? Press the issue? Demand preparations for a future no one wanted to imagine?

And if she did, would anyone listen?

The possibility that her warning might be ignored terrified her more than open refusal. Silence has doomed Eridani. Polite dismissal. Deferred action. The fatal comfort of believing there would be time later.

"There may be no later," she thinks. Her jaw tightens.

The weight of responsibility presses down on her, heavier than rank or medals. She could already feel doubt gnawing at the edges of her conviction, threatening to soften her resolve. But doubt, she reminds herself, is a luxury Eridani had paid for in blood.

"We must not wait for disaster to teach us vigilance," she tells herself, the words forming like a vow. "Justice, resilience, determination, unity—these are not ideals. They are defenses."

Grant straightens, pulling herself away from the ruined world below. Fear could not be allowed to paralyze her. History would not forgive hesitation.

Earth has to be warned.

Not just to prepare its weapons—but also its people. Its will. Its spirit.

Because if Earth follows the same path of denial and delay, it would not merely suffer Eridani's fate.

It would earn it.

Endnotes

1 Gabriel Soca for Gaia, "Pleiadians: Who They Are and What Their Mission on Earth Is," *The Pleiades: History and Meaning*, July 11, 2025, https://www.gaia.com/article/who-are-the-pleiadians.
2 Qin-Ra 2020, "Pleiadians are NOT aliens #2," https://www.pleiadianfamily.net/post/pleiadians-are-not-nordic-aliens.
3 Ibid.
4 Gabriel Soca for Gaia, "Pleiadians: Who They Are and What Their Mission on Earth Is," *The Pleiades: History and Meaning*, July 11, 2025, https://www.gaia.com/article/who-are-the-pleiadians.
5 World Population 2026, https://www.worldometers.info/world-population.
6 Youtube video, "I asked a captured alien about Why he fears the name Jesus," https://www.youtube.com/watch?v=Pqpvl17GjgY
7 Maharishi Mahesh Yogi, *Knowledge is Structured in Consciousness*."
8 Epsilon Eridani, https://www.thoughtco.com/closest-stars-to-earth-3073628.